Studies in Computational Intelligence

Volume 817

Series editor

Janusz Kacprzyk, Polish Academy of Sciences, Warsaw, Poland

The series "Studies in Computational Intelligence" (SCI) publishes new developments and advances in the various areas of computational intelligence—quickly and with a high quality. The intent is to cover the theory, applications, and design methods of computational intelligence, as embedded in the fields of engineering, computer science, physics and life sciences, as well as the methodologies behind them. The series contains monographs, lecture notes and edited volumes in computational intelligence spanning the areas of neural networks, connectionist systems, genetic algorithms, evolutionary computation, artificial intelligence, cellular automata, self-organizing systems, soft computing, fuzzy systems, and hybrid intelligent systems. Of particular value to both the contributors and the readership are the short publication timeframe and the world-wide distribution, which enable both wide and rapid dissemination of research output.

The books of this series are submitted to indexing to Web of Science, EI-Compendex, DBLP, SCOPUS, Google Scholar and Springerlink.

More information about this series at http://www.springer.com/series/7092

Wiktor B. Daszczuk

Integrated Model of Distributed Systems

Wiktor B. Daszczuk
Institute of Computer Science
Warsaw University of Technology
Warsaw, Poland

ISSN 1860-949X ISSN 1860-9503 (electronic)
Studies in Computational Intelligence
ISBN 978-3-030-12837-1 ISBN 978-3-030-12835-7 (eBook)
https://doi.org/10.1007/978-3-030-12835-7

Library of Congress Control Number: 2019930961

© Springer Nature Switzerland AG 2020
This work is subject to copyright. All rights are reserved by the Publisher, whether the whole or part of the material is concerned, specifically the rights of translation, reprinting, reuse of illustrations, recitation, broadcasting, reproduction on microfilms or in any other physical way, and transmission or information storage and retrieval, electronic adaptation, computer software, or by similar or dissimilar methodology now known or hereafter developed.
The use of general descriptive names, registered names, trademarks, service marks, etc. in this publication does not imply, even in the absence of a specific statement, that such names are exempt from the relevant protective laws and regulations and therefore free for general use.
The publisher, the authors and the editors are safe to assume that the advice and information in this book are believed to be true and accurate at the date of publication. Neither the publisher nor the authors or the editors give a warranty, express or implied, with respect to the material contained herein or for any errors or omissions that may have been made. The publisher remains neutral with regard to jurisdictional claims in published maps and institutional affiliations.

This Springer imprint is published by the registered company Springer Nature Switzerland AG
The registered company address is: Gewerbestrasse 11, 6330 Cham, Switzerland

I dedicate this monograph to my friend Jurek Mieścicki

pamięci Jurka

gdzie wędrował?
trawa lekko skoszona jego stopami
a gwiazdy tłumnie umierały
robiąc miejsce dla zmyślonego smoka
połykającego swój ogon
połykającego świat cały

mrówki jego myśli toczyły wieczną bitwę
nacierając pod sztandarem mądrości
który padał zadeptany następną myślą
a sen o pajęczynie pamięci
uświadamiał spisek przyzwyczajeń

podróżując aleją
wzdłuż żywopłotu sekund
wyciągał rękę
a małe zdarzenia
łaskotały
i raniły
przykuwały codziennością konieczności
koniecznością codzienności

chciał tylko minąć następny kamień
lecz ten nie wydawał się inny od poprzednich
ani bardziej szczęśliwy
ani bardziej mokry
ani bardziej egipski

szczęśliwa gwiazda spadła
robiąc wiele spustoszenia
zaśmiała się niecierpliwie
na pogrzebie jutra

in memory of Jurek

where he wandered?
grass mowed slightly by his feet
and the stars were dying in droves
making room for imaginary dragon
swallowing its tail
swallowing the whole world

ants of his thoughts took an eternal battle
assaulting under the banner of wisdom
which felt trampled by a next thought
and a dream of a spider's web of memory
made someone conscious of conspiracy habits

traveling an alley
along the hedge of seconds
he extended his hand
and small events
were tickling
and hurting
attracted an everyday necessity
a necessity of everyday life

he just wanted to get passed the next stone
but this did not seem different from previous ones
nor more happy
nor wetter
nor more Egyptian

a lucky star felt down
doing a lot of havoc
it laughed impatiently
at the funeral of the next day

Preface

Deadlock, a situation in which the system or part of it just stops and cannot continue successfully, is a typical error identified in distributed computer systems. However, verification of computer systems often requires some knowledge about the mathematical background and formal methods from the designer. In many institutions, formal verification is necessarily required, for example, in avionics or NASA projects. On the other hand, many engineers and students give up formal verification because time and effort must be sacrificed to study formalism. In many cases, they only decide to test, without formal proofs of proper cooperation in distributed systems.

The author offers original formalism—Integrated Model of Distributed Systems (IMDS)—combined with model checking based on temporal logic. The formalism allows the designer a specification that emphasizes the natural features of distributed systems and provides automatic verification of the typical features of such systems, such as deadlock and termination. The verified system is defined in a uniform model that can be observed in two general perspectives highlighting different aspects of the system. The server view involves servers' states and message exchange. The agent view presents distributed agents traveling by means of messages and collaborating through shared resources.

The introduced IMDS formalism exploits natural features of distributed systems:

- *Locality*: Individual actions that are part of the processes are executed based on the local server's state and the set of currently pending messages at this server, regardless of what is happening in other servers. No incident from outside the server—except for messages sent to it—can influence the operation of the server. Servers have no access to any global or non-local variables.
- *Autonomy*: Every decision about executing an action on the server—including the choice between many possible actions—is made autonomously. The only manner to influence the behavior of the server is to send a message to it, which may enable a previously impossible action.

- *Asynchrony*: Any required synchronization is hidden inside processes; therefore, processes are perceived as asynchronous. In addition, unidirectional communication channels between servers are naturally observed as asynchronous. Ergo, asynchrony is built into the grounds of IMDS.

Many modeling formalisms are based on synchronous actions performed by processes or on synchronous communication. In the author's opinion, asynchronous behavior is more realistic in the description of distributed systems, in which the only way to contact another server is to send a message and wait for response. The IMDS formalism emphasizes the enumerated characteristics of distributed systems: locality, autonomy and asynchrony.

The main research contribution of this monograph is the formulation of IMDS, in which communication duality is exposed, and which supports locality, autonomy and asynchrony in the specification of distributed systems.

Integration of IMDS with model checking allows detecting different types of deadlock and checking distributed termination using general formulas. These general formulas are not related to any specific structure of the model being verified. Communication deadlocks are observed in the server view, while resource deadlocks are observed in the agent view. This dichotomy is a rarity among commonly used formalisms. An important feature of IMDS is also the automatic distinction between deadlocks and distributed termination, in which processes stop deliberately.

General formulas for deadlock detection and termination checking enable automatic verification of these features. Many formalisms include automatic verification of total features, which involve all processes in the system. Partial deadlocks and partial termination—involving only a subset of all processes—must be individually determined by the designer in terms of features of the system under check. This requires the designer's knowledge, time and effort. Some formalisms support the design process by automatic detection of partial deadlocks and/or termination, but at the expense of limiting themselves to a certain class of verified systems, for example, only cycling systems. This monograph analyzes some of such approaches with their limitations. IMDS in conjunction with model checking is free from such restrictions, because a large class of systems can be automatically checked for partial and total deadlocks and termination. The obvious limitations of the introduced formalism are the finiteness of the model (for the purpose of static model checking) and the lack of broadcast communication (because messages are process carriers).

In automatic verification, temporal formulas specifying the features are generated by the Dedan program, developed by the author. The user does not need to know any temporal logic. The Dedan tool reports an error and provides a counterexample leading to a deadlock or a non-terminating loop. In this way, the designer can work in the style of "specify and push the button." The integration of IMDS with model checking and elaboration of general temporal formulas to verify deadlocks and termination, including their partial form, are both the research and the practical contribution of the author.

For the presented IMDS formalism, the textual specification language is developed and a graphical form of distributed automata is introduced. In addition, the conversion to equivalent Petri nets is defined. These three equivalent specification methods are used for various kinds of analysis:

- basic textual form for model checking and simulation over global reachability space,
- distributed automata for simulation over individual distributed components,
- Petri net for finding structural properties of the verified system, such as separation of components or unreachable actions; in Petri net analysis, many deadlocks may be found using single verification, while model checking finds only one of possibly many deadlocks.

The Dedan program, developed by the author, is based on the introduced methodology of specification in IMDS and on temporal verification. The program supports the specification of distributed systems, in textual or graphical form. A conversion from the server view of the system to the agent view is done algorithmically. Temporal verification is hidden inside the program; the user sees the results in an easily readable form of sequence-diagram-like counterexamples. The internal TempoRG verifier and external, commonly used temporal verifiers are used for model checking. In addition, a global reachability space of a checked system may be explored manually and its behavior may be observed in several simulation modes. Conversions of system specification to Petri net or to distributed automata model are supported. The translation of IMDS into Petri net and to distributed automata DA^3, invented by the author, and the development of the Dedan program are the practical contribution.

External verifiers used in Dedan do not support some kinds of fairness, which sometimes results in reporting false deadlocks. The author's original verification algorithm checking by spheres (CBS) is free from the mentioned disadvantage. The algorithm was improved for Dedan program using reverse reachability, which supports identification of individual cases in the reachability space.

The imposition of time constraints on the actions executed in the verified system, and on the message passing channels, may change the behavior of the system. For example, some deadlocks may disappear, while other deadlocks may arise. The timed version of IMDS formalism is presented (T-IMDS), with conversion rules to timed automata, in order to verify distributed systems with real-time constraints. This conversion preserves the mentioned features of distributed systems: locality, autonomy and asynchrony.

In order to verify very large systems whose reachability spaces cannot be built even under external, efficient verifiers, the non-exhaustive search algorithm "2-vagabonds" was developed. It is unique in partial deadlock and partial termination checking. The first vagabond puts hypotheses about deadlock/lack of termination, while the second vagabond verifies these hypotheses. The algorithm has a very limited set of parameters, in contrast to typical non-exhaustive algorithms.

The strong-fairness-aware algorithm, the timed model of distributed systems, preserving communication duality, locality, autonomy and asynchrony, and the non-exhaustive search algorithm for partial deadlock and termination checking are research and practical contributions.

Warsaw, Poland Wiktor B. Daszczuk

Acknowledgements

Stanisław Chrobot, Włodzimierz Zuberek and Jerzy Mieścicki helped to formulate the current version of Integrated Model of Distributed Systems. Janusz Sosnowski, Piotr Parewicz, Andrzej Pająk and Bogdan Czejdo gave important advice.

The following fragments are taken from the following publications (with major or minor changes):

- Chapter 2: (Daszczuk 2017) permission by Oxford University Press, Great Clarendon Street, Oxford, UK, OX2 6DP,
- Chapters 3 and 4, Figures in Chap. 1: (Daszczuk 2018) CC BY 4.0 license by MDPI, St. Alban-Anlage 66, 4052 Basel, Switzerland,
- Figure 5.9: (Czejdo et al. 2016) CC BY 4.0 license by Instutyt Wydawnicy Spatium Sp. z o.o., 25 Czerwca lok. 62, 26–600 Radom, Poland,
- Section 6.5: (Daszczuk 2019), permission license by Springer Nature,
- Figure 7.1, Sects. 7.2 and 7.3: (Daszczuk and Zuberek 2018), permission license by Springer Nature,
- Chapter 9: (Daszczuk and Zuberek 2018), permission license by Springer Nature,
- Figure 12.1—permission by Maria Daszczuk.

References

Czejdo, B., Bhattacharya, S. ... Daszczuk, W. B. (2016). Improving resilience of autonomous moving platforms by real-time analysis of their cooperation. *Autobusy-TEST*, *17*(6), 1294–1301. http://www.autobusy-test.com.pl/images/stories/Do_pobrania/2016/nr%206/logistyka/10_l_czejdo_bhattacharya_baszun_daszczuk.pdf

Daszczuk, W. B. (2017). Communication and resource deadlock analysis using IMDS formalism and model checking. *The Computer Journal*, *60*(5), 729–750. https://doi.org/10.1093/comjnl/bxw099

Daszczuk, W. B. (2018). Specification and verification in integrated model of distributed systems (IMDS). *MDPI Computers*, *7*(4), 1–26. https://doi.org/10.3390/computers7040065

Daszczuk, W. B. (2019). Asynchronous specification of production cell benchmark in integrated model of distributed systems. In R. Bembenik, L. Skonieczny, G. Protaziuk, M. Kryszkiewicz, & H. Rybinski (Eds.), *23rd International Symposium on Methodologies for Intelligent Systems, ISMIS 2017, Warsaw, Poland, 26–29 June 2017, Studies in Big Data, vol. 40* (pp. 115–129). Cham, Switzerland: Springer International Publishing. https://doi.org/10.1007/978-3-319-77604-0_9

Daszczuk, W. B., Zuberek, W. M. (2018). Deadlock detection in distributed systems using the IMDS formalism and petri nets. In W. Zamojski, J. Mazurkiewicz, J. Sugier, T. Walkowiak, & J. Kacprzyk (Eds.), *12th International Conference on Dependability and Complex Systems, DepCoS-RELCOMEX 2017, Brunów, Poland, 2–6 July 2017. AISC vol. 582* (pp. 118–130). Cham, Switzerland: Springer International Publishing. https://doi.org/10.1007/978-3-319-59415-6_12

Highlights

- IMDS formalism for specification of distributed systems, based on servers with their states and agents with their messages. Behavior modeled as actions over states and messages.
- Communication duality: The IMDS system decomposed into server processes communicating by messages or into agent processes communicating by servers' states. Two views of a system: server view and agent view.
- IMDS exposes natural features of distributed systems: locality of actions on servers, no global variables, autonomy of servers and asynchrony of processes and of communication.
- IMDS combined with model checking provides automatic verification of deadlocks: in communication or over resources, total and partial, and of distributed termination: total and partial.
- The Dedan program developed for specification, verification and simulation of distributed systems, based on IMDS formalism and model checking.
- Alternative formulation of IMDS in Petri nets for structural analysis: finding separate subsystems, unreachable actions and multiple deadlocks.
- Distributed automata DA^3, equivalent to IMDS—simulation over distributed components, preserved communication duality (two kinds of automata: server-based and agent-based).
- CBS, original verification algorithm, supporting strong fairness (compassion).
- Cooperation of Dedan with external model checkers: Spin, NuSMV, Timeless Uppaal.
- Real-time-constrained version of IMDS (T-IMDS) cooperating with external timed verifier Uppaal, translation of T-IMDS to Uppaal Timed Automata.
- 2-vagabonds: non-exhaustive verification algorithm for partial deadlock detection and partial termination checking.

Contents

Chapter 1
Introduction

1.1 Overview of the Formalism

The author's experience with building industrial and research distributed systems showed that proper modeling and verification of such systems is crucial for their quality (Schaefer and Hahnle 2011). Parallelism of tasks in a concurrent system creates difficulties due to the possibly infinite set of their common executions. In ICS, WUT, a verification methodology based of the original CSM [Concurrent State Machines, invented by Mieścicki (2006)] was developed and implemented in COSMA design tool (Mieścicki et al. 1996; Daszczuk 2000, 2001, 2002). The CSM is based on the principle that signals caused by the system component are apparent to other components as long as they are being generated. This feature can be treated as a kind of synchrony or a non-local state in the system: the internal state of the component is visible to the entire system. The modules can make their decisions based on the actual states of other modules.

The author built temporal verifier TempoRG (Daszczuk 2003) in the COSMA environment (Mieścicki 2006), and carried out many verification experiments. Parts of several real-life systems developed in ICS, WUT were subject to temporal verification in the COSMA tool by the author and his collaborators (Grabski et al. 1999; Daszczuk et al. 2001a, b; Mieścicki et al. 2003, 2004; Mieścicki and Daszczuk 2006):

- KURMAN distributed system for identification and on-line control of industrial plants (Mieścicki et al. 1992; Daszczuk et al. 1998a), implementing the original MIKOZ plant identification method (Mieścicki et al. 1995),
- ESS industrial plant monitoring distributed system (Daszczuk et al. 1995), for many years working in Polish energy plants, awarded Siemens Prize 1996; this system was later the basis for the research on distributed SCADA systems in inter-faculty project PATIA,

© Springer Nature Switzerland AG 2020

W. B. Daszczuk, *Integrated Model of Distributed Systems*, Studies in Computational Intelligence 817, https://doi.org/10.1007/978-3-030-12835-7_1

- remote monitoring protocols using SMS GSM messages and data mode (Mieścicki et al. 1998; Daszczuk et al. 1998b), in cooperation with DeTeMobil (currently T-Mobile),
- embodied mobile cooperating agents built by ICS, WUT students (Daszczuk and Mieścicki 2001).

The verification process identified several behavioral errors and communication errors in these systems.

Modern systems add a new difficulty dimension to parallelism: independent nodes cooperating in the distributed environment. The level of parallelism is much higher than in centralized systems, and there is no real global state of calculations in such a system. There are no global/non-local variables as well as no global state of the system exists. The only manner of communication between tasks is message passing over the network. The server cannot even expect that a response to a sent message would ever come With the exception of strictly related systems using the common bus, system components communicate with messages sent asynchronously.

Such an environment should be considered in modern systems of grid computing, mobile systems, distributed sensor networks, cloud computing, etc. A new Internet of Things (IoT) paradigm was introduced, in which independent, autonomous nodes communicate using simple protocols to agree their coordinated behavior (Lee et al. 2013).

The verification of many systems under COSMA, mentioned above, and many others, was successful, but the author realized that old-fashioned models based on synchronous actions in separate servers were not suitable for distributed systems. In a distributed system there is neither a global nor a non-local state, therefore no coordinated action between two or more servers can be performed synchronously. The server can inform another server about its state and its intentions by means of a message. Then the server can wait for the response to its request, again passed by a message. Each server operates autonomously, deciding on its individual behavior, observing local variables that store the server state value and contain messages pending locally. Asynchrony means that usually the server is waiting for a message, or the message pends waiting for acceptance by the server. Message passing is also asynchronous, based on sending messages along unidirectional asynchronous channels, and possible responses are passed through separate channels in the opposite direction.

The new IMDS (Integrated Model of Distributed Systems) formalism was proposed by the author and Stanisław Chrobot (Chrobot and Daszczuk 2006), then the research was continued by the author alone (Daszczuk 2008, 2017a). The IMDS is a closed model of a distributed system, i.e., it models resources and their users: no external influence is accepted. Resources and users are modeled as distributed servers, and distributed computations over the servers are modeled as traveling agents. The modeling of closed systems is a typical basis for verification of synchronization solutions, in particular the technique of static model checking (Clarke et al. 1999), used in this monograph. The IMDS formalism highlights the natural features of distributed systems, rarely found in other models: servers locality, autonomy of their decisions and asynchrony of performed actions and communication. However, the most important feature of IMDS is communication duality, which allows the observation of a uniformly specified distributed system in one of two perspectives: servers communicating by messages or agents communicating via the servers' states.

The IMDS formalism is founded on the basic observation of the server operation in a distributed environment, namely that the server remains in a certain state p, determined by the values of its internal variables (Fig. 1.1a). If the message m appears at the server in the context of an agent (Fig. 1.1b), it can be accepted, which means that the action invoked by this message is executed. The execution of the action is performed entirely in the context of this server (without any external intervention). As a result of the action, the server variables get new values, which means the next server state p' (Fig. 1.1c). In addition, another message m' is generated which continues the calculation represented by the agent (Fig. 1.1d). This simple scheme of the server and agent activity in the distributed environment leads to the definition of IMDS. Two basic sets are defined: *states* and *messages*. The set of *actions* is defined as the relation between the input pairs (*message, state*) and output pairs (*next message, next state*). The only exception to this rule is the agent-terminating action, in which only the new server state appears, but there is no next agent message on the output of the action (the action does not have the feature presented in Fig. 1.1d).

Note that the scheme presented in Fig. 1.1(b–d) is only an illustration to give the reader an intuition of the action. In IMDS, the execution of the action is instant and indivisible, with no phases shown in the figure.

The distributed system is defined over *servers* and *agents*—distributed computations performed in the system. Messages are passed between servers in the context of individual agents, just as the states are replaced by next states in the context of individual agents. This duality is the basis for the verification of behavioral features: communication deadlock, resource deadlock and distributed termination (Chap. 4).

The present formulation of IMDS, described in (Daszczuk 2018b) (Chap. 3), differs from previous articles (Chrobot and Daszczuk 2006; Daszczuk 2008; 2017a), where distributed systems were defined over four sets: servers, values of their states, services offered, and agents. Processes were defined as sets of states and

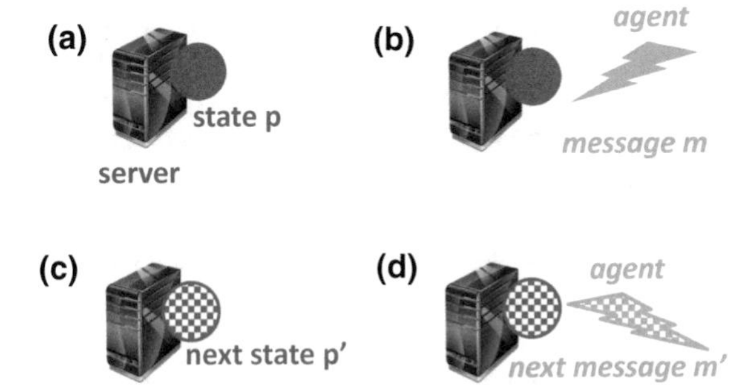

Fig. 1.1 The basic principle of IMDS: **a** the server and its state, **b** the message arrives, **c** the action is executed—the server changes its state, **d** the new message is issued

messages. In the present formulation, the basic concepts are states and messages, and the processes are defined as sets of actions. The new formulation is more convenient for readers. However, the two formulations are equivalent in the case for asynchronous sequential processes. Other types of processes can be defined in the previous formulation, but we do not deal with them in this monograph. In addition, in the papers (Chrobot and Daszczuk 2006; Daszczuk 2008) dynamic creation of servers and agents was allowed in actions, but for the purpose of static model checking we restrict the creation of servers and agents to the start of the system.

Communication in distributed systems is described in the literature in two basic models that are alternative to each other. A client-server model (Jia and Zhou 2005) shows independent servers, actively cooperating by message passing. Servers change their states upon messages received. Processes in such a model are identified with the servers.

On the other hand, a distributed system can be described in the paradigm of Remote Procedure Call (RPC), with processes traveling between the servers via procedure invocations and returns (Jia and Zhou 2005). Many such processes can originate on a single server, and they migrate to other servers treated passively as shared resources. This model can be extended to processes traveling through servers, without having to return to the server from which they originate.

The main research contribution of this monograph is the Integrated model of Distributed Systems which covers the two views in a uniform specification. A distributed system is defined as a set of actions of servers and agents. The two system views can be extracted from a single action-based specification, simply by grouping the servers' actions. This corresponds to the client-servers paradigm (Jia and Zhou 2005). The server view is obtained by grouping the actions of individual servers. Processes are identified with servers and communicate using messages. The agent view is retrieved by grouping actions of individual agents. Processes represent distributed computations (as in the RPC model) called agents. Agents collaborate by shared variables which are represented by servers' states. Combining two views - cooperating servers or migrating agents - into a uniform specification is defined by the author as "communication duality". The two concepts, servers and agents, can usually be assigned to observable system elements. For example in the automatic vehicle guidance system, described in Sect. 5.4 (Czejdo et al. 2016), servers are road segments controllers while agents are vehicles. In the production cell example [Sect. 5.6 (Daszczuk 2019)], servers are subcontrollers of the devices, and agents are identified with metal plates traveling through the cell.

There are several models of parallel execution in computer systems (Kessler and Keller 2007). They can be generally divided into synchronous and asynchronous models (Milner 1983; Savoiu et al. 2002; van Glabbeek et al. 2008). In synchronous models, cooperating processes (or threads, tasks or other activities running in parallel) must simultaneously communicate being in specific states. Therefore, at least cooperative processes must reconcile their states or, alternatively, participate in a required manner in the global state of the system. Examples of synchronous formalisms are: LOTOS (Rosa and Cunha 2004), Büchi automata (Holzmann 1995) used for verification of systems using linear temporal logic LTL (Clarke et al.

1999), Zielonka's automata (Zielonka 1987) or Timed Automata (Alur and Dill 1994). In each of these models, the automata synchronize using common symbols on transitions. On the other hand, CSP (Hoare 1978) and Uppaal timed automata (Behrmann et al. 2006) synchronization is performed by execution of simultaneous *send* and *receive* operations. Likewise, synchronous operations on complementary input and output *ports* in CCS (Milner 1984) and Occam (May 1983) require the agreement of the processes.

We argue that the above models are inadequate for modeling of distributed systems, because such systems do not have a global (or non-local) state. In reality, the processes cannot agree their states. The only way to influence the behavior of another server is sending a message to it. On the other hand, receiving a reply does not guarantee that the responding server is still in the state reported in the response. Therefore, asynchronous formalisms in which the processes do not agree on common states are more realistic in modeling of distributed systems (Johnsen et al. 2009).

Message Passing Automata [MPA (Bollig and Leucker 2004)] and Pushdown Distributed Automata [PDA (Balan 2009)] are really asynchronous, because the received messages are waiting for acceptance in specific structures: queues or stacks. The locality of actions is assured, because the acceptance of the message depends only on the local state of receiving automaton and the locally stored message: the one waiting for the longest time in a queue or, respectively, for the shortest time on a stack. However, in both formalisms, the automata are not fully autonomous, because they do not decide locally which message is accepted first (it is decided upon the underlying rule: FIFO or LIFO).

Many modeling environments use one of the mentioned communication paradigms: message passing in client-server paradigm or RPC paradigm with messages serving as process migration means. None of them provides duality of communication in a unified formalism. IMDS offers this duality as the two projections of a modeled system: on servers or on agents. The projections simply consist in grouping the actions: in individual servers or in individual agents. This is always possible because each action has a pair of state and message on input. This duality allows the expression of various features that manifest in individual views: communication deadlocks in the server view, resource deadlocks and distributed termination in the agent view (Chap. 4).

In IMDS, actions are executed on the server basing on its local state and on messages pending at it. This allows modeling of distributed systems in an asynchronous way, which is realistic and natural. We argue that IMDS is well-suited for modeling of distributed systems, because the actions in servers preserve:

- *locality* of the execution of actions, which depends only on the local state and messages pending locally; the server does not have access to a global or non-local state;
- *autonomy* of decisions: the local structure of the server determines which of the messages pending at the server will be accepted first, no order among the pending messages is assumed;

- *asynchrony of actions*: received messages are pending and they wait for acceptance; if there is no pending message or the server cannot accept any of the pending messages in the current state, then the server waits for the appropriate message; on the other hand, pending messages wait for the appropriate server state;
- *asynchrony of communication*: a message is sent along unidirectional asynchronous channel; the sender doesn't know when the message reaches the target server and when the response can be expected (if the calculation provides a response).

This integration of communication duality, locality, autonomy and asynchrony in a one model is unique among formalisms.

Intensive testing of computer systems can improve the reliability of the software, but it does not guarantee its correctness. In distributed systems, where parallelism is very intense, reachability space (and the number of possible behaviors) could be huge, and testing does not provide dependability. What's more, there is no global system state, which further aggravates verification. For such systems, formal proofs of correctness and formal verification are needed to create trustable programs. Formal models are needed to express the properties of distributed systems and to verify them (Schaefer and Hahnle 2011).

Model checking and other formal methods are widely and effectively used in the verification of computer systems, for example in avionic systems (Moy et al. 2013) or NASA projects (NASA 2007). Formal verification came even to everyday work in industry (Miller et al. 2010; Fahland et al. 2011). However, the author's experience with formal methods showed that the verification procedure should be automatized as much as possible, allowing designers to check their projects without any assistance of verification specialists. For this purpose, a formalism for automatic deadlock detection and termination checking was developed (Daszczuk 2017a). It combines IMDS with the model checking technique. In the author's opinion, the most important features are the deadlock freedom and the inevitable termination (in the case of non-cyclic systems). There are many automated deadlock and termination verification methods (these are listed in Chap. 2), but generally they focus on the total features, which involve all system's processes. On the other hand, partial features (regarding a subset of processes) can be checked in a number of methods, but only in the case of systems with a very limited structure. The second research contribution of the author is the combination of the two approaches: automated verification of partial (and total) features in systems having arbitrary shape, restricted only by the IMDS principle. This principle is the definition of a distributed system as a set of actions, which are the relation between state and message pairs (or a state only in the case of agent-terminating actions).

General temporal formulas, unrelated to individual system structures, were developed to detect deadlocks and check distributed termination. Communication duality allows expressing the features of a distributed system in two aspects: communication deadlocks in the server view and resource deadlocks and distributed termination in the agent view. Because the formulas concern the behavior of individual processes, partial deadlocks and termination (for a subset of the processes in the system) can be identified.

1.2 Modeling of Distributed Systems with Communication Duality

As mentioned in the previous section, the distributed system is usually described as Client-Server or as RPC model (Jia and Zhou 2005). In the former approach, the system is defined in terms of servers exchanging messages. A server is an entity that has a specific state and offers a number of actions. Execution of such an action changes the state of the server. The process in a system is defined as a sequence of server state changes. These changes are internal to the server processes, which communicate with other server processes via message exchange.

However, a second model is possible: if the process is associated with a traveling agent, it migrates between the servers and performs steps of its computations on various servers. The agent communicates with other traveling agents via servers' states. Messages are process carriers and therefore hidden in the processes. In this way, the system is described in terms of resource sharing instead of message passing. This is similar to a Remote Procedure Calling principle [RPC (Jia and Zhou 2005)[1]], although the traveling process model is wider, because generally the process need not return to its "home" server.

The key fact is that in both approaches, Client-Server and RPC, this is one and the same system, shown in one of two views, depending on how the actions are grouped. The action gives a detailed description of the operation in a distributed system: it is invoked by an incoming message, executed on the server and changes the state of this server. The assumption in our model is that each action (with the exception of agent-terminating actions) generates the next agent message after its completion. This message is issued to another server in the system to invoke the next action in the distributed computation. The server states and the agent messages are the input and output elements of the action. Therefore, there is an intermediate state or an intermediate message between a pair of subsequent actions. We may say that the state or the message "threads" two actions.

If actions in a sequence are threaded by server states—it is the server view. The server's states are the process carrier, while messages are the means of communication between servers. If the sequence is threaded by messages—it is the agent view. Agent messages are the process carrier, while servers' states serve to communicate agent processes. These two views are decompositions showing the aspects of message passing and resource sharing in a distributed system.

Figure 1.2 presents an example of a distributed system—a bounded buffer with capacity 2. The system consists of three servers: *buffer*, *producer* and *consumer*,

[1]Yet, in general it is not necessary for a process to return to the calling server.

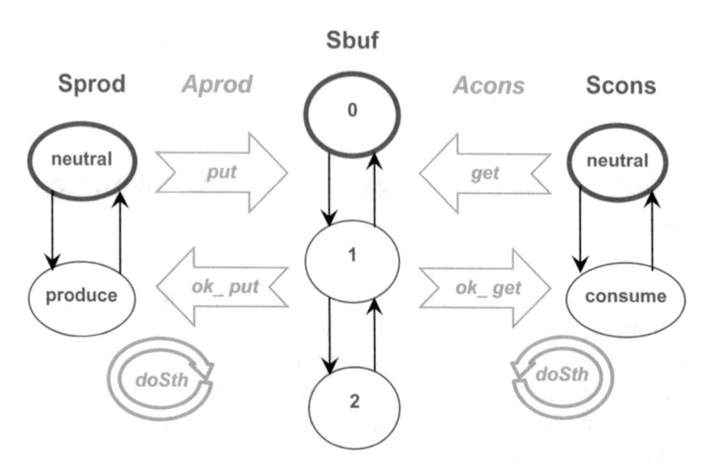

Fig. 1.2 "Bounded buffer" example: three servers *Sbuf*, *Sprod* and *Scons*, two agents *Aprod* and *Acons*

their names: `Sbuf, Sprod` and `Scons`. The `Sprod` server produces items and sends them to `Sbuf` using `put` message, then receives `ok_put` response message. The `Sprod` server sometimes does something else, which is modeled by `doSth` message sent to itself. The `Scons` server acts similar. The `Sbuf` server receives `put` and `get` messages, but accepts `put` only when the buffer is not full and accepts `get` only if the buffer is not empty. We can distinguish two agents representing distributed computations: `Aprod`, consisting of all messages sent and received by `Sprod`, and `Acons` - similarly. The servers' states are ovals and the agents' messages are contour arrows. Agent messages are shown in italics. System actions are shown as thin arrows between servers states. The figure does not explain which messages are accepted in which states, but this is obvious in such a simple system. For example, an empty buffer (`0` state) accepts the `put` message, changes its state to `1` and generates `ok_put` message. In the server view, the system is decomposed into three server processes, each of which groups the actions of a given server. In the agent view, the system is decomposed into two agent processes, each of which groups actions of a given computation (`producing` or `consuming`). Initial server states and initial agent messages have bold edges. In the figure (and all subsequent ones) the servers and their states are colored red, while agents and their messages are green.

Informally, IMDS is based on states, messages and actions. If the server state allows the action to be executed (we say that the state and the message *match*), the server can *fire* the action. The action changes the server state to another one and issues the next message, invoking an action on another server (or sometimes on the same server). The set of actions is the relation between the input pairs (m,p) and the output pairs (m', p'). States in the system are attributed to servers while messages are attributed to agents (formal definition is given in Chap. 3). The process is a sequence of actions: on the same server (*server process*) or in the same agent

(*agent process*). The system starts from the initial set of servers' states (one state for the server) and the initial set of agents' messages (one message for the agent).

There is a special type of actions for process termination in IMDS. In such an action, the output message does not exist and therefore the agent terminates. This type of actions is the relation between the pair (m,p) and singleton (p').

The obvious limitation of the modeled system class comes from the IMDS rules: because there is one input message and one output message in the regular action (no output message in the agent-terminating action), systems with broadcast communication cannot be specified. Instead, the designer can use a set of agents ("task force") to perform broadcast-like communication.

The behavior of the distributed system may change if specific time constraints are imposed on its elements. Many specification and verification models have time-dependent versions, such as timed Petri nets and Timed Automata (Bérard et al. 2005; Popescu and Martinez Lastra 2010). The author provides a timed version of IMDS (T-IMDS, with real-time constraints). The translation rules to Uppaal Timed Automata (Behrmann et al. 2006) are defined for verification purposes. The T-IMDS, described in Chap. 10, preserves the features needed to verify distributed systems: communication duality, locality, autonomy and asynchrony. T-IMDS is both a research and a practical contribution of the author.

1.3 Distributed Deadlock and Termination

The verification methodology, consisting of the IMDS specification and temporal verification (Daszczuk 2008), allows the designer to specify a distributed system, observe the server and agent views of the system and to easily find various types of deadlocks (total/partial deadlock, communication/resource deadlock) and distributed termination of processes (again total/partial).

An example of deadlock in the IMDS server view is when two servers are waiting for messages from each other. It is a communication deadlock, because the way the server processes cooperate is message passing. The definition of a communication deadlock in IMDS is a situation where there are pending messages on the server (at least one), but they can never be served. A deadlock can affect all servers (total deadlock) or a subset of servers (partial deadlock). A deadlock can even involve a single server.

In the agent view, deadlock is a situation where an agent has a message pending at some server, but the agent cannot make any progress. The message cannot be served, but messages of other agents possibly can. Again, the deadlock can involve all agents (total deadlock) or subset of agents (partial deadlock). Note that following this definition, a single process can be deadlocked. Such a situation is sometimes called *starvation* (Tai 1994) or *stall* (Masticola and Ryder 1990).

A (partial) termination in IMDS is simply the disappearance of the agent: the agent-terminating action binds the input pair (*message, state*) with the singleton output (*next state*). The action produces the next state of the server but no next message is generated.

Model checking is a method often used in static deadlock detection or termination verification in distributed systems (Clarke et al. 1999; Baier and Katoen 2008). The model checker automatically detects situations such as deadlock or lack of termination, using system specification and temporal formulas. General[2] temporal formulas for deadlock and termination are defined by the author over the elements of IMDS (Daszczuk 2017a, 2018b). The generality of formulas means that they do not depend on the specific features of processes in a given system. Only general features of server processes and agent processes are used as arguments. The two views of a model allow to find communication deadlocks and resource deadlocks. In other formalisms, there are general formulas to find a total deadlock (see Chap. 2), but most of them do not distinguish total deadlock from total termination. Some verifiers perform automatic detection of partial deadlocks, but only in cases where the scheme of the verified system is strictly limited. A similar limitation concerns checking partial termination. If the structure of the system is not restricted, the designer must have some knowledge of temporal logic or equivalent formalisms in order to perform the verification. In addition, none of the known verifiers distinguish between communication deadlocks and resource deadlocks.

If the process falls into a deadlock, or if it does not terminate, the model checker creates a counterexample for the program defect. The counterexample supports the designer in finding and correcting errors in the specification. A counterexample is a sequence of states and messages leading from an initial situation[3] to an erroneous situation.

General formulas—used to deadlock detection and termination checking—allow automatic verification. Therefore, the formulation and verification of the properties are carried out automatically inside the modeling program Dedan ("Dedan," n.d.), and the user only sees the final results. What the designer *must* absorb is the IMDS formalism and specification language, which is much simpler than the specification of temporal formulas.

Automatic verification of: partial and total deadlocks, partial and total distributed termination, and distinction between deadlock and termination are research and practical contribution of this monograph. They are the basis for the development of the Dedan program.

A non-exhaustive search algorithm, called "2-vagabonds", has also been developed to detect deadlocks and termination in very large reachability spaces.

---

[2]We call the formulas "general", because "universal formulas" usually refer to those containing the universal quantifier.

[3]Note that we do not use the term "state" to avoid ambiguity.

1.4 Practical Outcome

The Dedan environment was developed by the author for the practical application of the proposed verification methodology ("Dedan," n.d.), which is presented in Chaps. 4, 5 and 6. Further work of the author, described in this monograph, concerns additional facilities that support the specification and verification of distributed systems. First, IMDS model can be transformed into a Petri net (with a strictly restricted scheme) (Heiner and Heisel 1999) and verified by cooperation with Charlie Petri net analyzer (Heiner et al. 2015), described in Chap. 7. For this purpose, the translation of the IMDS specification into a Petri net is defined (Daszczuk and Zuberek 2018; Daszczuk 2017b).

Model checking is very attractive in deadlock detection, but its disadvantage results from its very nature: model checking is based on evaluation of the temporal formula. If a formula for "deadlock freeness" is evaluated as *false*, because a deadlock was found, the verifier stops. However, many other deadlocks can be hidden inside the verified system. Analysis of Petri net solves this problem: identification of bad siphons allows to find multiple deadlocks (Daszczuk and Zuberek 2018; Daszczuk 2018a). Temporal analysis allows for judging if a siphon represents a deadlock. What's more, Charlie can perform another structural analyzes, such as identifying separated subnets, which mean the existence of independent parts of the system (possibly due to the designer's error), and the location of unreachable transitions meaning "dead code" in the specification.

The author invented Asynchronous and Autonomous Distributed Automata [DA^3, Chap. 8 (Daszczuk 2017b)], which are equivalent to IMDS models, and which allow for graphical specification of the distributed system and the graphical simulation over its components. For the two views of a distributed system, server automata and agent automata are introduced. A counterexample obtained from model checking can be animated graphically in DA^3 automata during the examination of the error in the specification. This three equivalent specifications (IMDS, Petri net and DA^3) create a useful verification environment: IMDS for text specification, model checking (see Sect. 1.3) and simulation over the reachability space, DA^3 for graphical specification and simulation over distributed components, and Petri net for structural analysis. Reachability spaces of all three forms are equivalent.

The Dedan program includes an internal TempoRG model checker, which uses the original CBS algorithm, developed by the author (Daszczuk 2003). The verification algorithm is adapted to its use in the Dedan tool: it is simplified to a limited set of formulas (see Sect. 3.4), and reverse reachability is used to identify the distinct cases of the shape of the reachability space (Chap. 9). This verifier uses an explicit reachability space, which allows to deal with rather small specifications, such as student exercises. For larger specifications, external model checkers are used: Spin, NuSMV and Uppaal. Dedan's cooperation with external verifiers is effective, but some structures of verified systems give false positives (detection of non-existent deadlocks). It results from the lack of strong fairness (compassion) in their implementation. Problems with external verifiers and the construction of the CBS verification algorithm, which supports strong fairness, are described in Chap. 9. The presented three facilities: integration with Charlie, integration with external temporal verifiers and specification and simulation in DA^3, are practical contributions of the author (Daszczuk 2017b). What's more, the strong fairness-aware verification algorithm is both a research and a practical outcome.

The students of ICS, WUT prepared under the supervision of the author the Rybu preprocessor (Daszczuk et al. 2017), allowing the specification of systems in imperative programming style. This preprocessor is also described in the monograph, Sect. 5.8.

1.5 The Organization of the Monograph

Chapter 2 presents some techniques for detecting deadlocks and checking termination, and discusses their advantages and disadvantages. Chapter 3 introduces the IMDS formalism: the static description of the configuration of a distributed system according to states and messages, and the behavior of a distributed system in terms of actions. The semantics of a distributed system behavior is introduced as the *Labeled Transition System* [LTS (Reniers and Willemse 2011)] over the actions. Deadlock and termination are defined in terms of the LTS. This LTS is the basis for temporal model checking in the Dedan environment, described in Chap. 4. Examples of distributed systems and their verification in Dedan are presented in Chap. 5. The Dedan program itself is described in Chap. 6. Chapters 7 and 8 show alternative formulations of IMDS systems: specification in Petri net and in the original Distributed Automata DA^3. Chapter 9 presents the construction of the own CBS verification algorithm for the Dedan environment. A general scheme for evaluation of temporal formulas is described. The timed IMDS version (with real-time constraints imposed on communication channels, states and actions) is covered in Chap. 10. The own algorithm "2–vagabonds" for non-exhaustive deadlock and termination detection is presented in Chap. 11. The conclusions and future directions of development of IMDS and Dedan are discussed in Chap. 12.

References

Alur, R., & Dill, D. L. (1994). A theory of timed automata. *Theoretical Computer Science, 126*(2), 183–235. https://doi.org/10.1016/0304-3975(94)90010-8.

Balan, M. S. (2009). Serializing the parallelism in parallel communicating pushdown automata systems. *Electronic Proceedings in Theoretical Computer Science, 3,* 59–68. https://doi.org/10.4204/EPTCS.3.5.

Baier, C., & Katoen, J.-P. (2008). *Principles of model checking*. Cambridge, MA: MIT Press. ISBN: 9780262026499.

Behrmann, G., David, A., & Larsen, K. G. (2006). *A tutorial on Uppaal 4.0*. Aalborg, Denmark. URL: http://www.it.uu.se/research/group/darts/papers/texts/new-tutorial.pdf.

Bérard, B., Cassez, F., Haddad, S., Lime, D., & Roux, O. H. (2005). Comparison of the expressiveness of timed automata and time Petri nets. In *Third International Conference, FORMATS 2005*, Uppsala, Sweden, 26–28 September 2005 (pp. 211–225). Berlin Heidelberg: Springer. https://doi.org/10.1007/11603009_17.

Bollig, B., & Leucker, M. (2004). Message-passing automata are expressively equivalent to EMSO logic. In *15th International Conference CONCUR 2004—Concurrency Theory, London, UK, 31 August–3 September 2004* (pp. 146–160). Berlin Heidelberg: Springer. https://doi.org/10.1007/978-3-540-28644-8_10.

Chrobot, S., & Daszczuk, W. B. (2006). Communication dualism in distributed systems with Petri net interpretation. *Theoretical and Applied Informatics*, *18*(4), 261–278. URL: https://taai.iitis.pl/taai/article/view/250/.

Clarke, E. M., Grumberg, O., & Peled, D. (1999). *Model checking*. Cambridge, MA: MIT Press. ISBN: 0-262-03270-8.

Czejdo, B., Bhattacharya, S., Baszun, M., & Daszczuk, W. B. (2016). Improving resilience of autonomous moving platforms by real-time analysis of their cooperation. *Autobusy-TEST*, *17* (6), 1294–1301. URL: http://www.autobusy-test.com.pl/images/stories/Do_pobrania/2016/nr%206/logistyka/10_l_czejdo_bhattacharya_baszun_daszczuk.pdf.

Daszczuk, W. B. (2000). *State space reduction for reachability graph of CSM automata*. ICS WUT Research Report No 10/2000. https://doi.org/10.13140/RG.2.2.17223.91047.

Daszczuk, W. B. (2001). Evaluation of temporal formulas based on "Checking By Spheres." In *Proceedings Euromicro Symposium on Digital Systems Design,* Warsaw, Poland, 4–6 September 2001 (pp. 158–164). IEEE. https://doi.org/10.1109/dsd.2001.952267.

Daszczuk, W. B. (2002). *Critical trees: counterexamples in model checking of CSM systems using CBS algorithm*. ICS WUT Research Report No. 8/2002. https://doi.org/10.13140/RG.2.2.13228.16003.

Daszczuk, W. B. (2003). *Verification of temporal properties in concurrent systems*. Warsaw University of Technology. URL: https://repo.pw.edu.pl/docstore/download/WEiTI-0b7425b5-2375-417b-b0fa-b1f61aed0623/Daszczuk.pdf.

Daszczuk, W. B. (2008). *Deadlock and termination detection using IMDS formalism and model checking, Version 2*. ICS WUT Research Report No. 2/2008. https://doi.org/10.13140/RG.2.2.23294.48969.

Daszczuk, W. B. (2017a). Communication and resource deadlock analysis using IMDS formalism and model checking. *The Computer Journal*, *60*(5), 729–750. https://doi.org/10.1093/comjnl/bxw099.

Daszczuk, W. B. (2017b). Threefold analysis of distributed systems: IMDS, Petri net and distributed automata DA3. In *37th IEEE Software Engineering Workshop, Federated Conference on Computer Science and Information Systems, FEDCSIS'17, Prague*, Czech Republic, 3–6 September 2017 (pp. 377–386). IEEE Comput. Soc. Press. https://doi.org/10.15439/2017f32.

Daszczuk, W. B. (2018a). Siphon-based deadlock detection in Integrated Model of Distributed Systems (IMDS). In *Federated Conference on Computer Science and Information Systems, 3rd Workshop on Constraint Programming and Operation Research Applications (CPORA'18)*, Poznań, Poland, 9–12 September 2018 (pp. 421–431). IEEE. https://doi.org/10.15439/2018f114.

Daszczuk, W. B. (2018b). Specification and verification in integrated model of distributed systems (IMDS). *MDPI Computers*, *7*(4), 1–26. https://doi.org/10.3390/computers7040065.

Daszczuk, W. B. (2019). Asynchronous specification of production cell benchmark in integrated model of distributed systems. In R. Bembenik, L. Skonieczny, G. Protaziuk, M. Kryszkiewicz, & H. Rybinski (Eds.), *23rd International Symposium on Methodologies for Intelligent Systems, ISMIS 2017, Studies in Big Data*, Vol. 40, Warsaw, Poland, 26–29 June 2017 (pp. 115–129). Cham, Switzerland: Springer International Publishing. https://doi.org/10.1007/978-3-319-77604-0_9.

Daszczuk, W. B, & Mieścicki, J. (2001). *JADE—A platform for research on cooperation of physical and virtual agents*. ICS WUT Research Report No 15/2001.
Daszczuk, W. B., & Zuberek, W. M. (2018). Deadlock detection in distributed systems using the IMDS formalism and Petri nets. In W. Zamojski, J. Mazurkiewicz, J. Sugier, T. Walkowiak, & J. Kacprzyk (Eds.), *12th International Conference on Dependability and Complex Systems, DepCoS-RELCOMEX 2017*, Brunów, Poland, 2–6 July 2017. AISC (Vol. 582, pp. 118–130). Cham, Switzerland: Springer International Publishing. https://doi.org/10.1007/978-3-319-59415-6_12.
Daszczuk, W. B., Grabski, W., Wytrębowicz, J. (2001a). System modeling in the COSMA environment. In *Proceedings Euromicro Symposium on Digital Systems Design*, Warsaw, Poland, 4–6 September 2001 (pp. 152–157). Warsaw, Poland: IEEE Comput. Soc. https://doi.org/10.1109/dsd.2001.952264.
Daszczuk W B, Mieścicki J, Nowacki, M., & Wytrębowicz, J. (2001b). System level specification and verification using concurrent state machines and COSMA environment. In *8th International Conference on Mixed Design of Integrated Circuits and Systems, MIXDES'01*, Zakopane, Poland, 21–23 June 2001 (pp. 525–532). arXiv:1703.05541.
Daszczuk, W. B., Mieścicki, J., Lewandowski, J., & Świrski, K. (1995). System for the integration of the block digital control system with the company network (in Polish: System Integracji Blokowego Cyfrowego Systemu Kontroli i Sterowania z Siecią Zakładową). In *Problemy Badawcze Energetyki Cieplnej*, Warszawa, 5–8 Grudnia 1995 (p. tom 1, 101–107). Warszawa: Oficyna Wydawnicza Politechniki Warszawskiej.
Daszczuk, W. B., Bielecki, M., & Michalski, J. (2017). Rybu: Imperative-style preprocessor for verification of distributed systems in the Dedan environment. In *KKIO'17—Software Engineering Conference*, Rzeszów, Poland, 14–16 September 2017. Polish Information Processing Society. arXiv:1710.02722.
Daszczuk, W. B., Mieścicki, J., Kochan, A., Zając, K., Skurzak, K., & Duplicki, K. (1998a). *A system for identification of fluidized bed furnace at the institute of heat engineering, WUT* (In Polish: System identyfikacji paleniska fluidalnego w Instytucie Techniki Cieplnej PW). ICS WUT Research Report No 23/98. https://doi.org/10.13140/RG.2.2.14008.55049.
Daszczuk, W. B., Mieścicki, J., Krystosik, A., et al. (1998b). *The organization of mobile computing systems using the GSM network as the communication subsystem*. ICS WUT Technical Report No.15/1998.
Dedan. (n.d.). URL: http://staff.ii.pw.edu.pl/dedan/files/DedAn.zip.
Fahland, D., Favre, C., Koehler, J., Lohmann, N., Völzer, H., & Wolf, K. (2011). Analysis on demand: Instantaneous soundness checking of industrial business process models. *Data & Knowledge Engineering*, *70*(5), 448–466. https://doi.org/10.1016/j.datak.2011.01.004.
Grabski, W., Daszczuk, W. B., Mieścicki, J., & Dobrowolski, H. (1999). *Verification of the event protocol for establishing and disconnecting a connection in the ESS system* (in Polish: Weryfikacja zdarzeniowego protokołu nawiązywania i rozłączania połączenia w systemie ESS). ICS WUT Research Report No. 10/1999.
Heiner, M, & Heisel, M. (1999). Modeling safety-critical systems with Z and Petri nets. In M. Felici, K. Kanoun, & A. Pasquini (Eds.), *SAFECOMP '99 Proceedings of the 18th International Conference on Computer Safety, Reliability and Security*, Toulouse, France, 27–29 September 1999, LNCS (Vol. 1698, pp. 361–374). Berlin Heidelberg: Springer-Verlag. https://doi.org/10.1007/3-540-48249-0_31.
Heiner, M., Schwarick, M., & Wegener, J.-T. (2015). Charlie—An extensible Petri net analysis tool. In *36th International Conference, PETRI NETS 2015,* Brussels, Belgium, 21–26 June 2015 (pp. 200–211). Cham, Switzerland: Springer International Publishing. https://doi.org/10.1007/978-3-319-19488-2_10.
Hoare, C. A. R. (1978). Communicating sequential processes. *Communications of the ACM, 21*(8), 666–677. https://doi.org/10.1145/359576.359585.
Holzmann, G. J. (1995). Tutorial: Proving properties of concurrent systems with SPIN. In *6th International Conference on Concurrency Theory, CONCUR'95*, Philadelphia, PA, 21–24 August 1995 (pp. 453–455). Berlin Heidelberg: Springer-Verlag. https://doi.org/10.1007/3-540-60218-6_34.

Jia, W., & Zhou, W. (2005). *Distributed network systems. From concepts to implementations.* NETA 15, 513. New York: Springer. https://doi.org/10.1007/b102545.

Johnsen, E. B., Blanchette, J. C., Kyas, M., & Owe, O. (2009). Intra-object versus inter-object: Concurrency and reasoning in creol. *Electronic Notes in Theoretical Computer Science, 243*, 89–103. https://doi.org/10.1016/j.entcs.2009.07.007.

Kessler, C., & Keller, J. (2007). Models for parallel computing: Review and perspectives. In *PARS-Mitteilungen*, 13–29 December 2007 (pp. 13–29). URL: https://www.ida.liu.se/~chrke55/papers/modelsurvey.pdf.

Lee, G. M, Crespi, N., Choi, J. K., & Boussard, M. (2013). Internet of Things. In *Evolution of Telecommunication Services, LNCS 7768* (pp. 257–282). Berlin Heidelberg: Springer-Verlag. https://doi.org/10.1007/978-3-642-41569-2_13.

Masticola, S. P, & Ryder, B. G. (1990). Static infinite wait anomaly detection in polynomial time. In *1990 International Conference on Parallel Processing,* Urbana-Champaign, IL, 13–17 August 1990 (Vol. 2, pp. 78-87). University Park, PA: Pennsylvania State University Press. URL: https://rucore.libraries.rutgers.edu/rutgers-lib/57963/.

May, D. (1983). Occam. *ACM SIGPLAN Notices, 18*(4), 69–79. https://doi.org/10.1145/948176.948183.

Mieścicki, J. (2006). The use of model checking and the COSMA environment in the design of reactive systems. *Annales UMCS, Informatica Vol. AI, 4AI*, 244–253. http://dlibra.umcs.lublin.pl/dlibra/plain-content?id=17558.

Mieścicki, J., & Daszczuk, W. B. (2006). Behavioral and real-time verification of a pipeline in the COSMA environment. *Annales UMCS, Informatica Vol. AI, 4AI*, 254–265. https://journals.umcs.pl/ai/article/view/3061.

Mieścicki, J., Baszun, M., Daszczuk, W. B., & Czejdo, B. (1996). Verification of Concurrent Engineering Software Using CSM Models. In *2nd World Conference on Integrated Design and Process Technology*, Austin, TX, 1–4 December 1996 (pp. 322–330). Dallas, TX: SDPS. arXiv:1704.06351.

Mieścicki, J., Dobrowolski, H., Daszczuk, W. B. (2003). *Model checking of multi-agent system's communication protocols specified using the Z notation.* ICS WUT Research Report No. 17/2003.

Mieścicki, J., Czejdo, B., Daszczuk, W. B. (2004). Model checking in the COSMA environment as a support for the design of pipelined processing. In *European Congress on Computational Methods in Applied Sciences and Engineering ECCOMAS 2004*, Jyväskylä, Finland, 24–28 July 2004. Jyväskylä, Finland. arXiv:1705.04728.

Mieścicki, J., Daszczuk, W. B., … Adamiec, R. (1992). *KURMAN: The system supporting the design of multidimensional automated control systems* (in Polish KURMAN: System wspomagający projektowanie wielowymiarowych układów automatycznej regulacji). ICS WUT Research Report No 10/92.

Mieścicki, J., Flisiak, D., Czuchaj, A., & Daszczuk, W. B. (1998). *Specification and verification of SMS-based data stream protocol*, ICS WUT Research Report No 16/98.

Mieścicki, J., Kurman, K. J., & Daszczuk, W. B. (1995). *MIKOZ: The method for identification and coordination of feedback interactions in the regulation of continuous technological processes* (in Polish: MIKOZ: Metoda identyfikacji i koordynacji oddziaływań zwrotnych). ICS WUT Research Report No. 26/95.

Milner, R. (1983). Calculi for synchrony and asynchrony. *Theoretical Computer Science, 25*(3), 267–310. https://doi.org/10.1016/0304-3975(83)90114-7.

Milner, R. (1984). *A calculus of communicating systems.* Berlin: Springer-Verlag.

Miller, S. P., Whalen, M. W., & Cofer, D. D. (2010). Software model checking takes off. *Communications of the ACM, 53*(2), 58–64. https://doi.org/10.1145/1646353.1646372.

Moy, Y., Ledinot, E., Delseny, H., & Monate, B. (2013). Testing or formal verification: DO-178C alternatives and industrial experience. *IEEE Software, 30*(3), 50–57. https://doi.org/10.1109/ms.2013.43.

NASA. (2007). *NASA systems engineering handbook.* NASA. URL: http://www.acq.osd.mil/se/docs/NASA-SP-2007-6105-Rev-1-Final-31Dec2007.pdf.

Popescu, C., & Martinez Lastra, J. L. (2010). Formal methods in factory automation. In J. Silvestre-Blanes (Ed.), *Factory automation* (pp. 463–475). Rijeka, Croatia: InTech. https://doi.org/10.5772/9526.

Reniers, M. A, & Willemse, T. A. C. (2011). Folk theorems on the correspondence between state-based and event-based systems. In *37th Conference on Current Trends in Theory and Practice of Computer Science*, Nový Smokovec, Slovakia, 22–28 January 2011 (pp. 494–505). Berlin: Springer-Verlag. https://doi.org/10.1007/978-3-642-18381-2_41.

Rosa, N. S., & Cunha, P. R. F. (2004). A software architecture-based approach for formalising middleware behaviour. *Electronic Notes in Theoretical Computer Science, 108,* 39–51. https://doi.org/10.1016/j.entcs.2004.01.011.

Savoiu, N., Shukla, S. K., & Gupta, R. K. (2002). Automated concurrency re-assignment in high level system models for efficient system-level simulation. In *Proceedings 2002 Design, Automation and Test in Europe Conference and Exhibition,* Paris, France, 4–8 March 2002 (pp. 875–881). IEEE Comput. Soc. https://doi.org/10.1109/date.2002.998404.

Schaefer, I., & Hahnle, R. (2011). Formal methods in software product line engineering. *Computer, 44*(2), 82–85. https://doi.org/10.1109/MC.2011.47.

Tai, K. (1994). Definitions and detection of deadlock, livelock, and starvation in concurrent programs. In D. P. Agrawal (Ed.), *1994 International Conference on Parallel Processing (ICPP'94)*, Raleigh, NC, 15–19 August 1994 (pp. 69–72). Boca Raton: CRC Press. https://doi.org/10.1109/icpp.1994.84.

van Glabbeek, R., Goltz, U., & Schicke, J.-W. (2008). On synchronous and asynchronous interaction in distributed systems. In *33rd International Symposium, MFCS 2008*, Toruń, Poland, 25–29 August 2008 (pp. 16–35). Berlin Heidelberg: Springer. https://doi.org/10.1007/978-3-540-85238-4_2.

Zielonka, W. (1987). Notes on finite asynchronous automata. *RAIRO—Theoretical Informatics and Applications, 21*(2), 99–135. https://doi.org/10.1051/ita/1987210200991.

Chapter 2
Related Work on Deadlock and Termination Detection Techniques

Because many deadlock detection techniques evolved, even the concept of deadlock varies across papers. Some of them define the deadlock as a feature of the system model, as is often the case with Wait-for-Graph techniques or model checking (Kaveh and Emmerich 2001; Karatkevich and Grobelna 2014; Karacali et al. 2000; Godefroid and Wolper 1992; Corbett 1996; Geilen and Basten 2003). The better formulation is abstract, but the disadvantage of this approach is often the lack of precision, as in: "Deadlock is a situation … in which a system or a part of it remains indefinitely blocked and cannot terminate its task" (Longley and Shain 1986). This meaning of the formulation is close to ours (except for its imprecision), because it includes resource and communication deadlock, and total and partial deadlock (sometimes called a livelock, because some processes work).[1] In some papers the infinite waiting of a single process is called *stall* (Masticola and Ryder 1990), while mutual waiting of at least two processes is called deadlock. The latter definition comes from the formulation of the Coffman deadlock conditions: mutual exclusion, hold and wait, no pre-emption condition, and circular-wait (Coffman et al. 1971).

Sometimes a deadlock is related only to resources: "Deadlocks arise when members of a group of processes which hold resources are blocked indefinitely" (Isloor and Marsland 1980) or only communication: "A global state contains a deadlock error when all of the communication channels are empty and all of the entities' states are receive states, where a receive state is a state whose outgoing transitions are all receive ones" (Huang 1989).

For the purpose of this monograph, we define deadlock as a situation in which the process (defined in Sect. 3.3) waits for a condition than cannot be fulfilled. The total deadlock applies to all the processes in the system, while the partial deadlock leaves some processes running. Resource deadlock and communication deadlock are defined in terms of the model of the distributed system in Sect. 3.4. Intuitively,

[1]We prefer calling a *livelock* the situation in which the process itself checks for a condition in a loop infinitely.

© Springer Nature Switzerland AG 2020

W. B. Daszczuk, *Integrated Model of Distributed Systems*, Studies in Computational Intelligence 817, https://doi.org/10.1007/978-3-030-12835-7_2

resource deadlock occurs when the agent process is waiting for a server state that cannot be reached. Communication deadlock occurs when the server process waits infinitely for a message that would allow it to perform an action. Other messages may arrive, but none of them will result in an action execution.

The first attempts to detect deadlocks were made on the global reachability space of a centralized system (or rather its model) (Holt 1972) by analysis of a dependency graph called "Wait-for Graph" (WFG). This approach allowed to statically predict the risk of deadlock. Alternative methods are based on observation of system dynamically in selected "snapshots", which allows to discover deadlocks in run-time (Chandy and Lamport 1985). This is especially useful in systems in which global behavior is hard to predict, like in a set of independent user programs requesting shared resources.

The Wait-for-Graph approach has been transferred to distributed systems after the introduction of the local observations, because global state cannot be determined in such systems. Locality means that there are no concepts, such as synchronized passage of time or simultaneous activities. A global decision about the deadlock is made on the basis of independent local circumstances reported by the system components (Chandy et al. 1983; Elmagarmid 1986; Mitchell and Merritt 1984). The Wait-for-Graphs concept is still successfully used, especially for detecting deadlocks at runtime (Agarwal and Stoller 2006; Knapp 1987).

In cases covered by two views of the distributed system (server view and agent view), separate techniques are used: resource deadlocks (Elmagarmid 1986; Zhou et al. 2004) and communication deadlocks, without buffers (Hilbrich et al. 2009, 2013; Natarajan 1986; Hosseini and Haghighat 2005), or with finite or infinite buffers (Allen et al. 2007; Olson and Evans 2005). Resource deadlock, communication deadlock and their distinction are discussed in (Singhal 1989; Knapp 1987).

The algorithms dedicated for data bases typically use Wait-for Graphs (Choudhary et al. 1989; Yeung et al. 1994; Park and Scheuermann 1991; Knapp 1987). A special method for data bases uses variable values: *Wait-for*, *Held-by* and *Request-Q* (Grover and Kumar 2013).

Pulse is a Linux deadlock finder (Li et al. 2005), which creates "ghost" copies of waiting processes and allows them to operate in a limited environment. Restrictions concern the omission of activities that result in permanent results, e.g., writing to a file. If the ghost process fulfils the condition that another process is waiting for, it means deadlock. This approach is effective, but sometimes finds non-existent deadlocks or misses real deadlocks.

Dynamic methods identify deadlocks in distributed systems during run-time, but if no deadlock is reported, it does not guarantee the deadlock freeness. This is because the system analysis continues "until the current moment", and a deadlock may occur in the future.

Another investigated feature of distributed system is termination. Various distributed termination checking techniques evolved (Huang 1989; Mattern 1989; Peri and Mittal 2004; Zhou et al. 2004). The methods are based on the observation of some features of distributed processes or control over message traffic. Sometimes special elements of distributed processes are defined for termination detection.

Model checking techniques are based on the exploration of the global reachability space of a system, or part of it in some techniques (Clarke et al. 1999; Bensalem et al. 2011). The methods are based on the temporal verification of the reachability space. System activities are expressed in the form of local features of its components, and the global reachability space of the system is constructed. The features of system components are expressed as temporal logic formulas and verified by evaluating these formulas. Deadlock is one of most important features. Model checkers are often equipped with automatic deadlock detection procedures (Corbett 1996; Puhakka and Valmari 2000), for example in Spin and PathFinder (Holzmann 1995, 1997; Havelund and Pressburger 2000). Usually, a total deadlock is identified as "a state with no future", i.e., a strongly connected subgraph containing only one state: the deadlock itself (Kaveh 2001; Magee and Kramer 1999). Deadlock freeness is checked by the CTL temporal formula **AG EX** *true* (for any state a next state exists) (Cho et al. 2006; Royer 2001; Perna and George 2006; Mazuelo 2008; Bérard et al. 2001; Kokash and Arbab 2008; Arcaini et al. 2009). However, the total termination seems to be analogous state that has no future. In a cyclic system, where termination of processes is not expected, the above formula identifies a deadlock. In systems in which processes terminate their run, total deadlock should be distinguished from total termination. Therefore, the formula is refined to differentiate deadlock from termination in (Corbett, 1994): **AG** ((!*FINAL*) $\Longrightarrow$ **EX** *true*) (for each state there is a next state, except for total termination states). However, the general concept of "*FINAL*" is not given in the paper.

Similar results are obtained using CCS (Milner 1984), based on succession relation. CCS term is in deadlock if the relation analogous to ¬(**AG EX** *true*) is not met (De Francesco et al. 1998). This has the same disadvantage as model checking: it does not distinguish between deadlock and termination and only finds total deadlocks. Suchlike is the verification using CSP (Shi and Liu 2010) and evaluation of specific formulas which depend on component Finite State Machines (FSM) structures (Hiraishi 2000). The methods include graph-based, for example deadlock detection in statecharts (Kaveh 2001), sequence diagrams-based (Zhang et al. 2015). Other method are language-based, as verification of Promela specification in Spin (Holzmann 1995).

As previously indicated, model checking is usually used to find total deadlocks, i.e., states in which all processes are waiting, and no progress is possible. Temporal formulas can also be used to find partial deadlocks, in which some processes are involved in a deadlock, but other processes can continue. Generally, in the case of partial deadlock detection, temporal formulas are based on the structure of verified models, for example, they use *key states*, as in: **AG**(*LQ.Head.Occ* $\Longrightarrow$ **AF**(¬*LQ.Head.Occ*)) (Kern and Greenstreet 1999) (whenever the head of the link queue is occupied with a request, this will eventually be processed and the head will be released). This identifies deadlocks in individual processes. Similar methods are described in other papers (Havelund and Skakkebæk 1999; Chang and Jackson

2006; Beneš et al. 2008; Cordeiro 2010; Inverso et al. 2015; Arcile et al. 2016). The disadvantage of this approach is that temporal formulas need to be developed individually for each analyzed system, using its specific features.

Some other approaches to partial deadlock detection use general temporal formulas (i.e., formulas not related to the structure of the model being verified), but such general formulas require models to have specific properties. The Inspect checking tool (Yang et al. 2008), based on dynamic model verification, does not accept the cyclic reachability spaces. In pairwise model checking (Attie 2016) the general formula **AG EX** *true* is used for each pair of interrelated processes running in shared memory.

If the system is non-terminating (cyclical), the discontinuation of the process is obviously a deadlock (Baier and Katoen 2008). Conversely, another method can only be ascribed to terminating processes (Fahland et al. 2011). Some detection methods are used for specific systems architectures. For example, the WickedXmas approach uses the nodes communicating through the queues (Joosten et al. 2014). The specific method is designed for tree-like component architectures (Martens 2009). Our approach finds the discontinuation of individual processes, using general temporal formulas, and it works for any type of system: cyclic, terminating or hybrid - built of both kinds of processes. Termination, like deadlock, is a kind of process discontinuation (see methods for termination detecting below), but our method distinguishes deadlock and termination.

Another set of static methods applies to Petri nets. Some of them are based on the analysis of the reachability graph of the Petri net (Duri et al. 1994). The total deadlock (or termination) is a leaf in the reachability graph—there is no outgoing transition. Therefore, the analysis of reachability graph is similar to the model checking techniques and they are usually combined as a temporal analysis of the graph. It is difficult to distinguish a deadlock from distributed termination, so these methods are mainly addressed to infinitely looping systems (Guan et al. 2012).

Reachability analysis is usually based on the exploration of the entire reachability space. In many models the reachability space can be extremely large. The combinatorial space explosion problem concerns models in which the size of reachability space increases exponentially with the number of concurrent components. Partial order reduction methods provide a way to alleviate the explosion problem by using equivalences of certain sequences. There are methods based on the independence of transitions: for interleaving systems (Godefroid and Wolper 1992; Penczek et al. 2000; Gerth et al. 1999) and for coincidence systems (Daszczuk 2003). In the case of Petri nets, stubborn sets were invented which exploit commutativity over the sequences of transitions (Valmari 1994; Valmari and Hansen 2010). All these methods preserve deadlocks in reduction process.

Alternatively, structural analysis of Petri net can be used. Structural analysis determines properties of models on the basis of their structure, therefore it is not necessary to explore the reachability space. Structural analysis of deadlocks is based on subnets called siphons (Schmid and Best 1978; Chu and Xie 1997). It can be shown that if a model is deadlocked, the unmarked places constitute a siphon. Structural analysis of deadlocks systematically finds elementary siphons.

Elementary siphons are those from which other siphons are composed. After identifying the siphons, they are checked for unmarking possibility. Usually, linear programming is used for this purpose. If no elementary siphon can be found that can become emptied, the model is deadlock-free (Craig and Zuberek 2008).

Many algorithms for deadlock detection are proposed based on finding siphons in a Petri net (Chu and Xie 1997; Tricas et al. 2014; Węgrzyn et al. 2004). Furthermore, partial deadlock can be found in Petri net, using a method comparable to Wait-for-Graph [searching for processes waiting for resources held by other processes in a cycle (Tricas et al. 2005)]. Again, identification of siphons is helpful. Other methods use the conversion of a Petri net to another graph (Bertino et al. 1998).

Some methods use matrix operations, usually to calculate reachability. The Resource Allocation Graph (RAG) is represented by the adjacency matrix. The solution is based on the calculation the powers of the matrix (Ni et al. 2009; Leibfried 1989) or the reduction of some columns and rows (Nguyen et al. 2014). In Petri nets used for modeling manufacturing systems, the analysis of the machine-job incidence matrix is applied (Petrovic et al. 2009).

In the SCOOP model (Heußner et al. 2015), the source code of the program is converted to a graph in which cyclic-wait discovery is statically applied, similarly to Wait-for Graph analysis.

Model checking techniques usually interrupt the verification when an error is found (i.e., a value of a temporal formula is determined). As the effect, only one error is reported, and after the system revision, the next verification is needed. Methods based on finding Petri net siphons are better in this aspect, because all elementary siphons in a given net can be identified in a single run In turn, the strong point of model checking is counterexample generation: a trace leading from initial state to a deadlock state, suitable for analysis of deadlock causes.

The features of various deadlock detection techniques are summarized in Table 2.1.

In the detection of resource deadlocks, several semantic models are used (Knapp 1987). The simplest is *One–resource* model, in which a process can have at most one outstanding request for a single resource at a time. In the *AND* model, a process is permitted to request a set of resources. It is blocked until it is granted all the resources it has requested. An alternative model of resource requests is the *OR* model. A request for numerous resources is satisfied by granting any requested resource. The *AND-OR* model is a generalization of the two previous models, in which the requests may specify any combination of *AND* and *OR* in the resource request. The *k–out–of–n* model allows the specification of requests to obtain any k available resources out of a pool of size n. It is a generalization of the *AND-OR* model. In the most general model, no underlying structure of resource requests is assumed.

Because resources in IMDS are not directly included in formalism, each semantics can be modeled. The occupation of resources is modeled the servers' states, while granting of resources is modeled as the messages. Most natural is *OR* model, because there may be many states matching the agent's message, and any of

Table 2.1 Advantages and disadvantages of various deadlock detection techniques

Technique	Resource deadlocks	Communication deadlocks	Resource/comm. distinguishable	Reusable resources	Consumable resources	Total deadlocks	Partial deadlocks	Distinguishable from termination	Deadlock freeness guaranteed	Many deadlocks found
Dynamic WFG-resource	+	−	+	+	+	+	+	+	−	−
Dynamic WFG-communication	−	+	+	−	−	+	+	+	−	−
Pulse	+	+	−	+	+	+	+	+	−	−
Model checking **AG EX true**	+	+	−	+	−	+	−	−	+	−
Petri net siphons	+	+	−	+	+	+	+	−	+	+
IMDS and model checking	+	+	+	+	−	+	+	+	+	−

Column headings
"resource deadlocks"—resource deadlocks found
"communication deadlocks"—communication deadlocks found
"resource/comm. distinguishable"—resource/communication deadlocks distinguishable
"reusable resources"—deadlocks over reusable resources found
"consumable resources"—deadlocks over consumable resources found
"total deadlocks"—deadlocks with all processes involved
"partial deadlocks"—deadlocks with not all processes involved
"distinguishable from termination"—deadlock distinguishable from termination
"deadlock freeness guaranteed"—deadlock freeness guaranteed if no deadlock found
"many deadlocks found"—many (possibly all) deadlocks found by siphon analysis

these states enables the agent. Yet, a server can be constructed in such a way that its state informs of fulfilment of a complicated condition over resources, for example in *AND-OR* or *k–out–of–n* model.

In the analysis of distributed systems, two kinds of processes discontinuation are observed: an undesired lack of progress (deadlock), which is an error, and an expected stopping called process termination. Deadlock detection and termination checking methods must distinguish these two kinds of discontinuation (Sharma and Bhargava 1987), or simply prohibit one of them. This is the reason why many deadlock detection techniques are only addressed to endlessly looping systems in which there is no termination (Baier and Katoen 2008; Guan et al. 2012).

Just as in a case of deadlock detection, dynamic (runtime) methods of termination detection require some instrumentation of a system. Usually, messages are sent that report the states of particular processes and there is a mechanism to combine them into a global decision on distributed termination (Sharma and Bhargava 1987; Dijkstra and Scholten 1980; Kumar 1985). There are methods differing in instrumentation, dealing with failed processes or link failures, acceptance of temporary network partitioning (Mattern 1987; Matocha and Camp 1998; Chalopin et al. 2007; Kshemkalyani and Mukesh 2008). Static termination checking methods are based on observation of terminal states of individual processes. Model checking techniques are suitable for this purpose, using either model-specific formulas (Ma 2007) or general ones (Ray and Ray 2001). Construction of Counting Agent (Garanina and Bodin 2014) can be applied both dynamically and statically. Transition invariants (Podelski and Rybalchenko 2003) allow to check whether every execution that starts in the initial state is finite (terminating).

In the monograph, we propose the application of IMDS, which highlights locality of properties of verified system and exploits communication dualism (message passing/resource sharing). In conjunction with model checking, IMDS allows the expression and detection of a partial deadlock and the distinction between deadlock and distributed termination. Moreover, the new method allows distinguishing between a resource deadlock and a communication deadlock.

The comparison includes dynamic methods to provide a broad overview of deadlock detection and distributed termination methods. Our method is static and requires calculation of total reachability space. The novelty consists in specification of general temporal formulas over the IMDS model, for partial deadlocks and termination, and to distinguish deadlock from distributed termination and communication deadlocks from resource deadlocks. The formulas are general because they are not related to the structure of the system being verified. General formulas allow the designer to verify without knowledge of temporal logic. The Dedan program is based on this paradigm and allows verification in the "push the button" style, because model checking techniques are hidden inside the Dedan tool.

References

Agarwal, R., & Stoller, S. D. (2006). Run-time detection of potential deadlocks for programs with locks, semaphores, and condition variables. In *Proceedings of the Workshop on Parallel and Distributed Systems: Testing and Debugging (PADTAD-IV), ISSTA, 2006, Portland, ME, 17–20 July 2006* (pp. 51–59). New York, NY: ACM. https://doi.org/10.1145/1147403.1147413.

Allen, G. E., Zucknick, P. E., & Evans, B. L. (2007). A distributed deadlock detection and resolution algorithm for process networks. In *2007 IEEE International Conference on Acoustics, Speech and Signal Processing—ICASSP '07, 15–20 April 2007, Honolulu, HI* (Vol.2, pp. II-33–II-36). New York, NY: IEEE. https://doi.org/10.1109/icassp.2007.366165.

Arcaini, P., Gargantini, A., & Riccobene, E. (2009). *AsmetaSMV: A model checker for AsmetaL models—Tutorial*. Url: https://air.unimi.it/retrieve/handle/2434/69105/96882/Tutorial_AsmetaSMV.pdf.

Arcile, J., Czachórski, T., … Rataj, A. (2016). Modelling and analysing mixed reality applications. In A. Gruca, A. Brachman, S. Kozielski, & T. Czachórski (Eds.), *Man-Machine Interactions 4: 4th International Conference on Man-Machine Interactions, ICMMI 2015 Kocierz Pass, Poland, 6–9 Oct. 2015* (pp. 3–17). Cham, Switzerland: Springer International Publishing. https://doi.org/10.1007/978-3-319-23437-3_1.

Attie, P. C. (2016). Synthesis of large dynamic concurrent programs from dynamic specifications. *Formal Methods in System Design, 47*(131), 1–54. https://doi.org/10.1007/s10703-016-0252-9.

Baier, C., & Katoen, J.-P. (2008). *Principles of model checking*. Cambridge, MA: MIT Press. ISBN: 9780262026499.

Beneš, N., Černá, I., … Zimmerova, B. (2008). A case study in parallel verification of component-based systems. *Electronic Notes in Theoretical Computer Science, 220*(2), 67–83. https://doi.org/10.1016/j.entcs.2008.11.014.

Bensalem, S., Griesmayer, A., … Peled, D. (2011). Efficient deadlock detection for concurrent systems. In *9th ACM/IEEE International Conference on Formal Methods and Models for Codesign, MEMOCODE 2011, Cambridge, UK, 11–13 July 2011* (pp. 119–129). New York, NY: IEEE. https://doi.org/10.1109/memcod.2011.5970518.

Bérard, B., Bidoit, M., … Schnoebelen, P. (2001). *Systems and software verification. Model-checking techniques and tools*. Heidelberg: Springer. ISBN: 978-3-642-07478-3, https://doi.org/10.1007/978-3-662-04558-9.

Bertino, E., Chiola, G., & Mancini, L. V. (1998). Deadlock detection in the face of transaction and data dependencies. In *19th International Conference ICATPN'98, Lisbon, Portugal, 22–26 June 1998* (pp. 266–285). Heidelberg: Springer. https://doi.org/10.1007/3-540-69108-1_15.

Chalopin, J., Godard, E., … Tel, G. (2007). About the termination detection in the asynchronous message passing model. In *33rd Conference on Current Trends in Theory and Practice of Computer Science, Harrachov, Czech Republic, 20–26 Jan. 2007* (pp. 200–211). Heidelberg: Springer. https://doi.org/10.1007/978-3-540-69507-3_16.

Chandy, K. M., & Lamport, L. (1985). Distributed snapshots: Determining global states of distributed systems. *ACM Transactions on Computer Systems, 3*(1), 63–75. https://doi.org/10.1145/214451.214456.

Chandy, K. M., Misra, J., & Haas, L. M. (1983). Distributed deadlock detection. *ACM Transactions on Computer Systems, 1*(2), 144–156. https://doi.org/10.1145/357360.357365.

Chang, F. S.-H., & Jackson, D. (2006). Symbolic model checking of declarative relational models. In *Proceeding of the 28th International Conference on Software Engineering—ICSE '06, Shanghai, China, 20–28 May 2006* (pp. 312–320). New York, USA: ACM Press. https://doi.org/10.1145/1134285.1134329.

Cho, J., Yoo, J., & Cha, S. (2006). NuEditor—A tool suite for specification and verification of NuSCR. In *SERA 2004: Software Engineering Research and Applications, Los Angeles, CA, 5–7 May 2004, LNCS Vol. 3647* (pp. 19–28). Heidelberg: Springer. https://doi.org/10.1007/11668855_2.

Choudhary, A. N., Kohler, W. H., … Towsley, D. (1989). A modified priority based probe algorithm for distributed deadlock detection and resolution. *IEEE Transactions on Software Engineering*, *15*(1), 10–17. https://doi.org/10.1109/32.21721.

Chu, F., & Xie, X.-L. (1997). Deadlock analysis of Petri nets using siphons and mathematical programming. *IEEE Transactions on Robotics and Automation, 13*(6), 793–804. https://doi.org/10.1109/70.650158.

Clarke, E. M., Grumberg, O., & Peled, D. (1999). *Model checking*. Cambridge, MA: MIT Press. ISBN: 0-262-03270-8.

Coffman, E. G., Elphick, M., & Shoshani, A. (1971). System deadlocks. *ACM Computing Surveys, 3*(2), 67–78. https://doi.org/10.1145/356586.356588.

Corbett, J. C. (1994). An empirical evaluation of three methods for deadlock analysis of Ada tasking programs. In *Proceedings of the 1994 International Symposium on Software Testing and Analysis—ISSTA '94,), Seattle, WA, 17–19 Aug. 1994* (pp. 204–215). New York, NY: ACM Press. https://doi.org/10.1145/186258.187206.

Corbett, J. C. (1996). Evaluating deadlock detection methods for concurrent software. *IEEE Transactions on Software Engineering, 22*(3), 161–180. https://doi.org/10.1109/32.489078.

Cordeiro, L. (2010). SMT-based bounded model checking for multi-threaded software in embedded systems. In *Proceedings of the 32nd ACM/IEEE International Conference on Software Engineering—ICSE '10, Cape Town, RSA, 2–8 May 2010* (Vol. 2, pp. 373–376). New York, NY: ACM Press. https://doi.org/10.1145/1810295.1810396.

Craig, D. C., & Zuberek, W. M. (2008). Two-stage siphon-based deadlock detection in Petri nets. In P. Petratos & P. Dandapami (Eds.), *Current advances in computing, engineering and information technology* (pp. 317–330). Palermo, Italy: Int. Society for Advanced Research.

Daszczuk, W. B. (2003). *Verification of temporal properties in concurrent systems*. Warsaw University of Technology. Url: https://repo.pw.edu.pl/docstore/download/WEiTI-0b7425b5-2375-417b-b0fa-b1f61aed0623/Daszczuk.pdf.

De Francesco, N., Santone, A., & Vaglini, G. (1998). State space reduction by non-standard semantics for deadlock analysis. *Science of Computer Programming, 30*(3), 309–338. https://doi.org/10.1016/s0167-6423(97)00017-8.

Dijkstra, E. W., & Scholten, C. S. (1980). Termination detection for diffusing computations. *Information Processing Letters, 11*(1), 1–4. https://doi.org/10.1016/0020-0190(80)90021-6.

Duri, S., Buy, U., … Shatz, S. M. (1994). Application and experimental evaluation of state space reduction methods for deadlock analysis in Ada. *ACM Transactions on Software Engineering and Methodology*, *3*(4), 340–380. https://doi.org/10.1145/201024.201038.

Elmagarmid, A. K. (1986). A survey of distributed deadlock detection algorithms. *ACM SIGMOD Record, 15*(3), 37–45. https://doi.org/10.1145/15833.15837.

Fahland, D., Favre, C., … Wolf, K. (2011). Analysis on demand: Instantaneous soundness checking of industrial business process models. *Data & Knowledge Engineering*, *70*(5), 448–466. https://doi.org/10.1016/j.datak.2011.01.004.

Garanina, N. O., & Bodin, E. V. (2014). Distributed termination detection by counting agent. In *23th International Workshop on Concurrency, Specification and Programming, CS&P, Chemnitz, Germany, 29 Sept.–1 Oct. 2014* (pp. 69–79). Berlin, Germany: Humboldt University. Url: http://ceur-ws.org/Vol-1269/paper69.pdf.

Geilen, M., & Basten, T. (2003). Requirements on the execution of Kahn process networks. In *ESOP'03 the 12th European Symposium on Programming, Warsaw, Poland, 7–11 April 2003, LNCS 2618* (pp. 319–334). Heidelberg: Springer. https://doi.org/10.1007/3-540-36575-3_22.

Gerth, R., Kuiper, R., … Szreter, M. (1999). Partial order reductions preserving simulations. In *Concurrency Specification and Programming (CS&P), Warsaw, Poland, 28–30 Sept. 1999* (pp. 153–171). Url: http://www.ipipan.waw.pl/~penczek/WPenczek/papersPS/IPI843-97.ps.gz.

Godefroid, P., & Wolper, P. (1992). Using partial orders for the efficient verification of deadlock freedom and safety properties. In *3rd International Workshop, CAV '91, Aalborg, Denmark, 1–4 July, 1991, LNCS 575* (pp. 332–342). Heidelberg: Springer. https://doi.org/10.1007/3-540-55179-4_32.

Grover, H., & Kumar, S. (2013). Analysis of deadlock detection and resolution techniques in distributed database environment. *International Journal of Computer Engineering & Science, 2* (1), 17–25. Url: www.ijces.org/media/3Iss-3-IJCES150449.pdf.

Guan, X., Li, Y., … Wang, S. (2012). A literature review of deadlock prevention policy based on petri nets for automated manufacturing systems. *International Journal of Digital Content Technology and Its Applications, 6*(21), 426–433. http://www.globalcis.org/jdcta/ppl/JDCTA2027PPL.pdf.

Havelund, K., & Pressburger, T. (2000). Model checking JAVA programs using JAVA PathFinder. *International Journal on Software Tools for Technology Transfer (STTT), 2*(4), 366–381. https://doi.org/10.1007/s100090050043.

Havelund, K., & Skakkebæk, J. U. (1999). Applying model checking in java verification. In *5th and 6th International SPIN Workshops, Trento, Italy, July 5, 1999, Toulouse, France, 21 and 24 Sept. 1999* (pp. 216–231). London, UK: Springer. https://doi.org/10.1007/3-540-48234-2_17.

Heußner, A., Poskitt, C. M., … Morandi, B. (2015). Towards practical graph-based verification for an object-oriented concurrency model. *Electronic Proceedings in Theoretical Computer Science, 181*, 32–47. https://doi.org/10.4204/eptcs.181.3.

Hilbrich, T., de Supinski, B. R., … Müller, M. S. (2009). A graph based approach for MPI deadlock detection. In *Proceedings of the 23rd International Conference on Supercomputing—ICS '09, Yorktown Heights, NY, 8–12 June 2009* (pp. 296–305). New York, NY: ACM Press. https://doi.org/10.1145/1542275.1542319.

Hilbrich, T., de Supinski, B. R., … Müller, M. S. (2013). Distributed wait state tracking for runtime MPI deadlock detection. In *Proceedings of the International Conference for High Performance Computing, Networking, Storage and Analysis on SC '13, Denver, CO, 17–21 Nov. 2013* (pp. 1–12). New York: ACM Press. https://doi.org/10.1145/2503210.2503237.

Hiraishi, H. (2000). Verification of deadlock free property of high level robot control. In *Proceedings of the Ninth Asian Test Symposium ATS 2000, Taipei, 4–6 Dec. 2000* (pp. 198–203). New York, NY: IEEE Comput. Soc. https://doi.org/10.1109/ats.2000.893625.

Holt, R. C. (1972). Some deadlock properties of computer systems. *ACM Computing Surveys, 4* (3), 179–196. https://doi.org/10.1145/356603.356607.

Holzmann, G. J. (1995). Tutorial: Proving properties of concurrent systems with SPIN. In *6th International Conference on Concurrency Theory, CONCUR'95, Philadelphia, PA, 21–24 Aug. 1995* (pp. 453–455). Heidelberg: Springer. https://doi.org/10.1007/3-540-60218-6_34.

Holzmann, G. J. (1997). The model checker SPIN. *IEEE Transactions on Software Engineering, 23*(5), 279–295. https://doi.org/10.1109/32.588521.

Hosseini, R., & Haghighat, A. T. (2005). An improved algorithm for deadlock detection and resolution in mobile agent systems. In *International Conference on Computational Intelligence for Modelling, Control and Automation and International Conference on Intelligent Agents, Web Technologies and Internet Commerce (CIMCA-IAWTIC'06), Vienna, Austria, 28–30 Nov. 2005,* (Vol. 2, pp. 1037–1042). New York, NY: IEEE. https://doi.org/10.1109/cimca.2005.1631606.

Huang, S.-T. (1989). Detecting termination of distributed computations by external agents. In *[1989] Proceedings. The 9th International Conference on Distributed Computing Systems, Newport Beach, CA, 5–9 June 1989* (pp. 79–84). New York, NY: IEEE Comput. Soc. Press. https://doi.org/10.1109/icdcs.1989.37933.

Inverso, O., Nguyen, T. L., … Parlato, G. (2015). Lazy-CSeq: A context-bounded model checking tool for multi-threaded C-programs. In *2015 30th IEEE/ACM International Conference on Automated Software Engineering (ASE), Lincoln, NE, 9–13 Nov. 2015* (pp. 807–812). New York, NY: IEEE. https://doi.org/10.1109/ase.2015.108.

Isloor, S. S., & Marsland, T. A. (1980). The deadlock problem: An overview. *Computer, 13*(9), 58–78. https://doi.org/10.1109/mc.1980.1653786.

Joosten, S. J. C., Julien, F. V., & Schmaltz, J. (2014). WickedXmas: Designing and verifying on-chip communication fabrics. In *3rd International Workshop on Design and Implementation of Formal Tools and Systems, DIFTS'14, Lausanne, Switzerland, 20 Oct. 2014* (pp. 1–8). Eindhoven, The Netherlands: Technische Universiteit Eindhoven. Url: https://pure.tue.nl/ws/files/3916267/889737443709527.pdf.

Karacali, B., Tai, K.-C., & Vouk, M. A. (2000). Deadlock detection of EFSMs using simultaneous reachability analysis. In *Proceeding International Conference on Dependable Systems and Networks. DSN 2000, New York, NY, 25–28 June 2000* (pp. 315–324). New York, NY: IEEE Comput. Soc. https://doi.org/10.1109/icdsn.2000.857555.

Karatkevich, A., & Grobelna, I. (2014). Deadlock detection in Petri nets: One trace for one deadlock? In *2014 7th International Conference on Human System Interactions (HSI), Costa da Caparica, Lisbon, Portugal, 16–18 June 2014* (pp. 227–231). New York, NY: IEEE. https://doi.org/10.1109/hsi.2014.6860480.

Kaveh, N. (2001). Using model checking to detect deadlocks in distributed object systems. In W. Emmerich & S. Tai (Eds.), *2nd International Workshop on Distributed Objects, Davis, CA, 2–3 Nov. 2000, LNCS vol. 1999* (pp. 116–128). London, UK: Springer. https://doi.org/10.1007/3-540-45254-0_11.

Kaveh, N., & Emmerich, W. (2001). Deadlock detection in distribution object systems. In *Proceedings of the 8th European Software Engineering Conference held Jointly with 9th ACM SIGSOFT International Symposium on Foundations of Software Engineering—ESEC/FSE-9, Vienna, Austria, 10–14 Sept. 2001* (pp. 44–51). New York, NY: ACM Press. https://doi.org/10.1145/503209.503216.

Kern, C., & Greenstreet, M. R. (1999). Formal verification in hardware design: A survey. *ACM Transactions on Design Automation of Electronic Systems, 4*(2), 123–193. https://doi.org/10.1145/307988.307989.

Knapp, E. (1987). Deadlock detection in distributed databases. *ACM Computing Surveys, 19*(4), 303–328. https://doi.org/10.1145/45075.46163.

Kokash, N., & Arbab, F. (2008). Formal behavioral modeling and compliance analysis for service-oriented systems. In F. S. de Boer (Ed.), *Formal Methods for Components and Objects - 7th International Symposium, FMCO 2008, Sophia Antipolis, France, 21–23 Oct. 2008, LNCS 5751* (pp. 21–41). Heidelberg: Springer. https://doi.org/10.1007/978-3-642-04167-9_2.

Kshemkalyani, A. D., & Mukesh, S. (2008). *Distributed computing. Principles, algorithms, and systems*. Cambridge, UK: Cambridge University Press. ISBN: 978-0-521-87634-6.

Kumar, D. (1985). A class of termination detection algorithms for distributed computations. In *Fifth Conference on Foundations of Software Technology and Theoretical Computer Science, New Delhi, India 16–18 Dec. 1985* (pp. 73–100). Heidelberg: Springer. https://doi.org/10.1007/3-540-16042-6_4.

Leibfried, T. F. (1989). A deadlock detection and recovery algorithm using the formalism of a directed graph matrix. *ACM SIGOPS Operating Systems Review, 23*(2), 45–55. https://doi.org/10.1145/858344.858348.

Li, T., Ellis, C. S., … Sorin, D. J. (2005). Pulse : A dynamic deadlock detection mechanism using speculative execution. In *USENIX Annual Technical Conference, Anaheim, CA, 10–15 April 2005* (pp. 31–44). Berkeley, CA: USENIX Association. Url: http://usenix.org/legacy/publications/library/proceedings/usenix05/tech/general/full_papers/li/li.pdf.

Longley, D., & Shain, M. (1986). *Dictionary of information technology*. London: Macmillan Press. ISBN: 0-19-520519-7.

Ma, G. (2007). *Model checking support for CoreASM: Model checking distributed abstract state machines using Spin, MSc Thesis*. Simon Fraser University, Burnaby, Canada. Url: http://summit.sfu.ca/system/files/iritems1/8056/etd2938.pdf.

Magee, J., & Kramer, J. (1999). *Concurrency: Models and programs—From finite state models to java programs*. Chichester: Wiley. ISBN: 0470093552.

Martens, M. (2009). *Establishing properties of interaction systems, PhD Thesis*. University of Mannheim. Url: http://ub-madoc.bib.uni-mannheim.de/2629/1/DrArbeitMyThesis.pdf.

Masticola, S. P., & Ryder, B. G. (1990). Static infinite wait anomaly detection in polynomial time. In *1990 International Conference on Parallel Processing, Urbana-Champaign, IL, 13–17 Aug. 1990* (Vol. 2, pp. 78–87). University Park, PA: Pennsylvania State University Press. Url: https://rucore.libraries.rutgers.edu/rutgers-lib/57963/.

Matocha, J., & Camp, T. (1998). A taxonomy of distributed termination detection algorithms. *Journal of Systems and Software, 43*(3), 207–221. https://doi.org/10.1016/s0164-1212(98)10034-1.

Mattern, F. (1987). Algorithms for distributed termination detection. *Distributed Computing, 2*(3), 161–175. https://doi.org/10.1007/bf01782776.

Mattern, F. (1989). Global quiescence detection based on credit distribution and recovery. *Information Processing Letters, 30*(4), 195–200. https://doi.org/10.1016/0020-0190(89)90212-3.

Mazuelo, C. L. (2008). *Automatic model checking of UML models*. Universitat Berlin. Url: http://www.iam.unibe.ch/tilpub/2008/lar08.pdf.

Milner, R. (1984). *A calculus of communicating systems*. Heidelberg: Springer LNCS vol. 92, https://doi.org/10.1007/3-540-10235-3.

Mitchell, D. P., & Merritt, M. J. (1984). A distributed algorithm for deadlock detection and resolution. In *Proceedings of the Third Annual ACM Symposium on Principles of Distributed Computing—PODC '84, Vancouver, Canada, 27–29 Aug. 1984* (pp. 282–284). New York, NY: ACM Press. https://doi.org/10.1145/800222.806755.

Natarajan, N. (1986). A distributed scheme for detecting communication deadlocks. *IEEE Transactions on Software Engineering, SE-12*(4), 531–537. https://doi.org/10.1109/tse.1986.6312900.

Nguyen, H. H. C., Le, V. S., & Nguyen, T. T. (2014). Algorithmic approach to deadlock detection for resource allocation in heterogeneous platforms. In *2014 International Conference on Smart Computing, Hong Kong, 3–5 Nov. 2014* (pp. 97–103). New York, NY: IEEE. https://doi.org/10.1109/smartcomp.2014.7043845.

Ni, Q., Sun, W., & Ma, S. (2009). Deadlock detection based on resource allocation graph. In *2009 Fifth International Conference on Information Assurance and Security, Xian, China, 18–20 Aug. 2009* (pp. 135–138). IEEE. https://doi.org/10.1109/ias.2009.64.

Olson, A. G., & Evans, B. L. (2005). Deadlock detection for distributed process networks. In *ICASSP '05. IEEE International Conference on Acoustics, Speech, and Signal Processing, Philadelphia, PA, 18–23 March 2005* (Vol. 5, pp. 73–76). New York, NY: IEEE. https://doi.org/10.1109/icassp.2005.1416243.

Park, Y. C., & Scheuermann, P. (1991). A deadlock detection and resolution algorithm for sequential transaction processing with multiple lock modes. In *1991 The Fifteenth Annual International Computer Software & Applications Conference, Tokyo, Japan, 11–13 Sept. 1991* (pp. 70–77). New York, NY: IEEE Comput. Soc. Press. https://doi.org/10.1109/cmpsac.1991.170154.

Penczek, W., Szreter, M., … Kuiper, R. (2000). Improving partial order reductions for universal branching time properties. *Fundamenta Informaticae, 43*(1–4), 245–267. https://content.iospress.com/articles/fundamenta-informaticae/fi43-1-4-13.

Peri, S., & Mittal, N. (2004). On termination detection in an asynchronous distributed system. In *17th ISCA International Conference on Parallel and Distributed Computing Systems (PDCS), San Francisco, CA, 15–17 Sept. 2004* (pp. 209–215). Url: http://www.iith.ac.in/~sathya_p/index_files/PDCS-Transtd.pdf.

Perna, J., & George, C. (2006). *Model checking RAISE specifications*. Url: http://i.unu.edu/media/unu.edu/publication/1550/report331.pdf.

Petrovic, T., Bogdan, S., & Sindicic, I. (2009). Determination of circular waits in multiple-reentrant flowlines based on machine-job incidence matrix. In *European Control Conference (ECC) 2009, Budapest, Hungary, 23–26 Aug. 2009* (pp. 4463–4468). New York, NY: IEEE. DOI: https://doi.org/10.23919/ECC.2009.7075103.

Podelski, A., & Rybalchenko, A. (2003). *Software model checking of liveness properties via transition invariants*, Research Report MPI–I–2003–2–00, Max Planck Institite fur Informatik, December 2003. Url: https://pure.mpg.de/pubman/faces/ViewItemOverviewPage.jsp?itemId=item_1819221.

Puhakka, A., & Valmari, A. (2000). Livelocks, fairness and protocol verification. In *16th World Conference on Software: Theory and Practice, Beijing, China, 21–25 Aug. 2000, IFIP* (pp. 471–479). Laxenburg, Austria: International Federation for Information Processing. Url: http://www.cs.tut.fi/ohj/VARG/publications/00-3.ps.

Ray, I., & Ray, I. (2001). Detecting termination of active database rules using symbolic model checking. In *5th East European Conference on Advances in Databases and Information*

Systems, ADBIS '01, Vilnius, Lithuania, 25–28 Sept. 2001 (pp. 266–279). London, UK: Springer. https://doi.org/10.1007/3-540-44803-9_21.

Royer, J. (2001). Formal specification and temporal proof techniques for mixed systems. In *Proceedings 15th International Parallel and Distributed Processing Symposium. IPDPS 2001, San Francisco, CA, 23–27 April 2001* (pp. 1542–1551). New York, NY: IEEE Comput. Soc. https://doi.org/10.1109/ipdps.2001.925139.

Schmid, H. A., & Best, E. (1978). *A step towards a solution of the liveness problem in Petri nets.* Newcastle upon Tyne. Url: http://www.cs.ncl.ac.uk/publications/trs/papers/114.pdf.

Sharma, N. K., & Bhargava, B. (1987). *A robust distributed termination detection algorithm,* Report 87-726. West Lafayette, IN. Url: http://docs.lib.purdue.edu/cgi/viewcontent.cgi?article=1626&context=cstech.

Shi, L., & Liu, Y. (2010). Modeling and verification of transmission protocols: A case study on CSMA/CD protocol. In *SSIRI-C 2010—4th IEEE International Conference on Secure Software Integration and Reliability Improvement Companion, Singapore, 9-11 June 2010* (pp. 143–149). Washington: IEEE Computer Society. https://doi.org/10.1109/ssiri-c.2010.33.

Singhal, M. (1989). Deadlock detection in distributed systems. *Computer, 22*(11), 37–48. https://doi.org/10.1109/2.43525.

Tricas, F., Colom, J. M., & Merelo, J. J. (2014). Computing minimal siphons in Petri Net models of resource allocation systems: An evolutionary approach. In *CEUR Workshop Proceedings CEUR Workshop Proceedings, Petri Nets and Software Engineering, Tunis, Tunisia, 23–24 June 2014* (Vol. 1160, pp. 307–322). Tunis. https://doi.org/10.1109/tsmca.2005.855751.

Tricas F, Garcia-Valles, F., … Ezpeleta, J. (2005). A Petri net structure—based deadlock prevention solution for sequential resource allocation systems. In *2005 IEEE International Conference on Robotics and Automation, Barcelona, Spain, 18–22 April 2005* (pp. 271–277). New York, NY: IEEE. https://doi.org/10.1109/robot.2005.1570131.

Valmari, A. (1994). State of the art report: STUBBORN SETS. *Petri Net Newsletter, 46,* 6–14. Url: www.cs.tut.fi/ohj/VARG/publications/94-1.ps.

Valmari, A., & Hansen, H. (2010). Can stubborn sets be optimal? In *31st International Conference, PETRI NETS 2010, Braga, Portugal, 21–25 June 2010* (pp. 43–62). Heidelberg: Springer. https://doi.org/10.1007/978-3-642-13675-7_5.

Węgrzyn, A., Karatkevich, A., & Bieganowski, J. (2004). Detection of deadlocks and traps in petri nets by means of Thelen's prime implicant method. *International Journal of Applied Math and Computer Science, 14*(1), 113–121. Url: https://www.amcs.uz.zgora.pl/?action=download&pdf=AMCS_2004_14_1_13.pdf.

Yang, Y., Chen, X., & Gopalakrishnan, G. (2008). *Inspect: A runtime model checker for multithreaded C programs, Report UUCS-08-004.* Salt Lake City, UT. Url: http://www.cs.utah.edu/docs/techreports/2008/pdf/UUCS-08-004.pdf.

Yeung, C.-F., Huang, S.-L., … Law, C.-K. (1994). A new distributed deadlock detection algorithm for distributed database systems. In *TENCON'94—1994 IEEE Region 10's 9th Annual International Conference on: "Frontiers of Computer Technology", Singapore, 22–26 Aug. 1994* (Vol. 1, pp. 506–510). IEEE. https://doi.org/10.1109/tencon.1994.369249.

Zhang, H., Aoki, T., & Chiba, Y. (2015). A spin-based approach for checking OSEK/VDX applications. In C. Artho & P. C. Ölveczky (Eds.), *FTSCS 2014: Formal Techniques for Safety-Critical Systems, Luxembourg, 6–7 Nov. 2014* (pp. 239–255). Cham, Switzerland: Springer. https://doi.org/10.1007/978-3-319-17581-2_16.

Zhou, J., Chen, X., … Chen, D. (2004). M-Guard: A new distributed deadlock detection algorithm based on mobile agent technology. In *2nd International Conference on Parallel and Distributed Processing and Applications ISPA'04, Hong Kong, China, 13–15 Dec. 2004* (pp. 75–84). Heidelberg: Springer. https://doi.org/10.1007/978-3-540-30566-8_13.

Chapter 3
Integrated Model of Distributed Systems

The IMDS formalism (Chrobot and Daszczuk 2006; Daszczuk 2008) is designed to describe closed systems, i.e., systems in which the behavior does not depend upon external events. Static model verification, aimed at verifying correctness, applies only to closed systems.

A distributed system is composed of *servers* and *agents*. Servers model nodes of a distributed system, while agents model distributed computations. Servers execute *actions* which are requested by agents. The requests have the form of messages. It is assumed that messages are immediately available to target servers (without looking into technical aspects of message delivery). When an action requested by a message is executed (on a server), the server changes its state and the agent issues the next message. Alternatively—when there is no message generated—the agent terminates.

3.1 Basic IMDS Definition

The essence of the IMDS approach is to describe the system using two sets and a binary relation on the Cartesian product of these two sets. The sets are:

$$\begin{aligned} P &= \{p_1, p_2, \ldots\} - \text{the finite set of } \textit{states} \text{ (of } \textit{servers}\text{)} \\ M &= \{m_1, m_2, \ldots\} - \text{the finite set of } \textit{messages} \text{ (of } \textit{agents}\text{)} \end{aligned} \quad (3.1)$$

while the action relation Λ is a binary relation on $M \times P$:

$$\Lambda \subset (M \times P) \times (M \times P) - \text{the set of } \textit{actions} \quad (3.2)$$

For the elements of Λ we use the notation $\lambda \in \Lambda$, $\lambda = ((m,p),(m',p')) = (m,p)\Lambda(m',p')$.

© Springer Nature Switzerland AG 2020

W. B. Daszczuk, *Integrated Model of Distributed Systems*, Studies in Computational Intelligence 817, https://doi.org/10.1007/978-3-030-12835-7_3

The execution of the action (requested in the message) changes the state of the server and issues another message (by the agent). Thus, the action execution transforms the current state and the pending message into the next state and the next message.

In the action $(m,p)\Lambda(m',p')$ we say that:

- the pair (m,p) *match*;
- (m,p) is the *input pair*;
- (m',p') is the *output pair*;
- the state p is *current*;
- the message m is *pending* (we use also a term *current message*);
- p' is the *next state*;
- m' is the *next message*.

The definition must be supplemented by the initial sets of states and messages:

$$\begin{aligned} &P_{ini} \subset P\text{—the set of } \textit{initial states} \\ &M_{ini} \subset M\text{—the set of } \textit{initial messages} \end{aligned} \tag{3.3}$$

In the example of the system presented in Fig. 1.2, the sets and the relation are:

P = {*p_neutral, produce, c_neutral, consume, 0, 1, 2*}
M = {*p_doSth, put, ok_put, c_doSth, get, ok_get*}
Λ = {((*p_doSth, p_neutral*),(*p_doSth, p_neutral*)), ((*p_doSth, p_neutral*),(*put, produce*)), ((*ok_put, produce*),(*p_doSth, p_neutral*)), ((*put, 0*),(*ok_put, 1*)), ((*put, 1*),(*ok_put, 2*)), ((*c_doSth, c_neutral*),(*c_doSth, c_neutral*)), ((*c_doSth, c_neutral*), (*get, consume*)), ((*ok_get, consume*),(*c_doSth, c_neutral*)), ((*get, 1*),(*ok_get, 0*)), ((*get, 2*),(*ok_get, 1*))}
P_{ini}= {*p_neutral, c_neutral, 0*}
M_{ini}= {*p_doSth, c_doSth*}.

The equal names of states and messages used in more than one server are preceded by "*p_*" for producer and by "*c_*" for consumer.

3.2 IMDS System Behavior

The behavior of a system described in IMDS is represented by a Labeled Transition Systems (LTS) (Reniers and Willemse 2011), i.e., a rooted labeled directed graph.

For LTS representing the behavior of an IMDS system, the nodes are system configurations[1] (sets of current states and pending messages), the root is the initial

[1]The term "state" is reserved for elements of the set P.

configuration and the set of labels is the set of actions of the system. The system starts with P_{ini} and M_{ini}, therefore the initial configuration is a union of them. The states and messages together are called items. The set of directed arcs is determined by the actions, which convert their *input configurations* to their *output configurations.*

$$
\begin{aligned}
&H = P \cup M\text{–the set of } \textit{items}\\
&T_{ini} = P_{ini} \cup M_{ini}\text{–the } \textit{initial configuration}\\
&T \subset H\text{–the } \textit{configuration}
\end{aligned}
\tag{3.4}
$$

The rule of obtaining $T_{out}(\lambda)$ from $T_{inp}(\lambda)$ for the action λ is:

$$\forall_{\lambda \in \Lambda} \lambda = ((m,p),(m',p'))\, T_{inp}(\lambda) \supset \{m,p\}, T_{out}(\lambda) = T_{inp}(\lambda) \backslash \{m,p\} \cup \{m',p'\} \tag{3.5}$$

$$\text{LTS} = \langle N, n_0, W \rangle, \tag{3.6}$$

where:

- N is the set nodes (configurations $\{T_0, T_1, \ldots\}$, $T_{ini} = T_0$);
- n_0 is the root, $n_0 \in N$ (configuration T_{ini});
- W is the set of directed labeled transitions, $W \subset N \times \Lambda \times N$, $W = \{(T_{inp}(\lambda), \lambda, T_{out}(\lambda)) \mid \lambda \in \Lambda\}$.

As single actions are LTS labels, this means the interleaving semantics of the system (Penczek et al. 2000), i.e., one action is executed at a time.[2] The execution of the action transforms the input configuration to the output configuration in such a way that the input pair (*message, state*) is removed and replaced by the output pair (*next message, next state*). The items (messages and states) not included in the input pair remain unchanged. The LTS contains all possible runs of the system as paths in the graph.

A fragment of LTS for the example in Fig. 1.2 is presented in Fig. 3.1.

3.3 IMDS Processes

The system starts from the initial states of all servers (one state for each server) and the initial messages of all agents (one message for each agent). In the action, the input message must be directed to the server appointed by the input state, because the action is executed on the server. The next state generated in the action appoints

[2]Other semantics is also available, for example "maximum concurrency" semantics in which multiple possible actions are fired at a time.

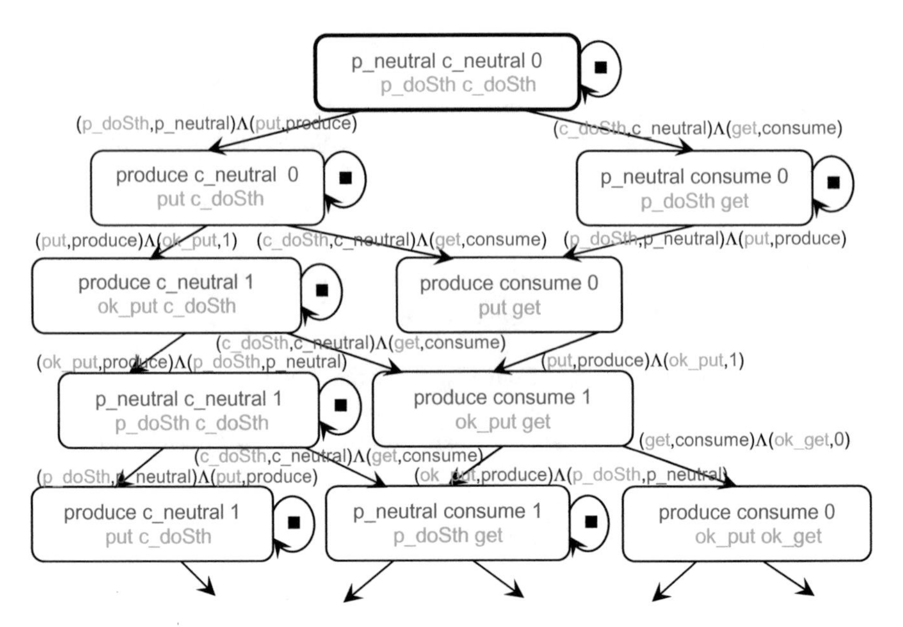

Fig. 3.1 A fragment of LTS for "Bounded buffer": configurations in nodes, actions on transitions. Filled square—self-loops with action *(c_doSth,c_neutral)Λ(c_doSth,c_neutral)* or *(p_doSth, p_neutral)Λ(p_doSth,p_neutral)*

the same server as the input state. This ensures continuity of the server behavior. Likewise, the next message generated in the action appoints the same agent as the input message. This ensures continuity of the agent behavior.

The formal definition of IMDS for the purpose of processes extraction follows:

$$
\begin{aligned}
&S = \{s_1, s_2, \ldots\} - \text{the finite set of servers}\\
&A = \{a_1, a_2, \ldots\} - \text{the finite set of agents}\\
&P(s) = \{p_{1s}, p_{2s}, \ldots\}, s \in S - \text{the finite set of states of server } s\\
&M(a) = \{m_{1a}, m_{2a}, \ldots\}, a \in A - \text{the finite set of messages of agent } a\\
&M(s) = \{m_{1s}, m_{2s}, \ldots\}, s \in S - \text{the finite set of messages directed to server } s
\end{aligned}
\tag{3.7}
$$

For the elements of sets *P*(*s*), *M*(*a*) and *M*(*s*), we say that these elements *appoint* corresponding server, agent and server, respectively.

The sets in the example are (initial elements are highlighted by boldface):

S = {*Sprod, Scons, Sbuf*}
A = {*Aprod, Acons*}
P(*Sprod*) = {***p_neutral***, *produce*}, *P*(*Scons*) = {***c_neutral***, *consume*}, *P*(*Sbuf*) = {***0***, *1*, *2*}

M(Aprod) = {***p_doSth***, *put*, *ok_put*}, *M(Acons)* = {***c_doSth***, *get*, *ok_get*}
M(Sprod) = {*p_doSth*, *ok_put*}, *M(Scons)* = {*c_doSth*, *ok_get*}, *M(Sbuf)* = {*put*, *get*}.

The constraints on *P*, *M* and *Λ*, needed for process extraction, are the following:

$$P = \cup_{s\in S}P(s), \forall_{s1,s2\in S}s_1 \neq s_2 \Rightarrow P(s_1)\cap P(s_2) = \varnothing$$
–every state is attributed to some server

$$M = \cup_{a\in A}M(a) = \cup_{s\in S}M(s),$$
$$\forall_{a1,a2\in A}a_1 \neq a_2 \Rightarrow M(a_1)\cap M(a_2) = \varnothing, \forall_{s1,s2\in S}s_1 \neq s_2 \Rightarrow M(s_1)\cap M(s_2) = \varnothing \quad (3.8)$$
–every message is attributed to some agent and is directed to some server.

The constraints on input and output pairs of the action are: the input state and input message concern the same server; the input and output states concern the same sever, the input and output messages concern the same agent:

$$\forall_{\lambda\in\Lambda}\lambda=((m,p),(m',p')), m\in M(a)\Rightarrow m'\in M(a), p\in P(s)\Rightarrow p'\in P(s)\wedge m\in M(s). \quad (3.9)$$

$\forall_{s\in S}$ *card*$(P(s)\cap P_{ini}) = 1$—the initial set of states contains exactly one state for every server
$\forall_{a\in A}$ *card*$(M(a)\cap M_{ini}) = 1$—the initial set of messages contains exactly one message for every agent.

We add a new type of agent-terminating action $(m,p)\Lambda(p')$, in which the next message is absent:
$\Lambda \subset (M \times P) \times (M \times P) \cup (M \times P) \times (P)$—the set of actions[3].

The constraints on the input pair and the output singleton in an agent-terminating action are:

$$\forall_{\lambda\in\Lambda}\lambda = ((m,p),(p')), p \in P(s) \Rightarrow p' \in P(s) \wedge m \in M(s). \quad (3.10)$$

The rule of obtaining $T_{out}(\lambda)$ from $T_{inp}(\lambda)$ for an agent-terminating action is:

$$\forall_{\lambda\in\Lambda}\lambda = ((m,p),(p')), T_{inp}(\lambda) \supset \{m,p\}, T_{out}(\lambda) = T_{inp}(\lambda)\setminus\{m,p\}\cup\{p'\}. \quad (3.11)$$

The definition of T_{ini} together with rules of obtaining T_{out} from T_{inp}, ensure that every configuration contains exactly one state for every server and at most one message for every agent (except for terminated agents).

Asynchronous sequential processes in the system are defined as sequences of actions. Note that the two output items of each action (or one output item only for

[3]Thus, *Λ* is not strictly a relation, because it contains both quadruples and triples. More formally, a message "agent termination" may be added to *M*, which is prohibited on input of any action.

the agent-terminating action) can be input items of other actions. Thus a sequence of actions is threaded by intermediate items. If we choose states as a process carrier—we get the *server process* running along with the changing server states. The states between actions define the *succession relation* between actions. In this perspective, intermediate messages are passed between actions of distinct processes: they serve as communication means between servers. Note that the server processes do not terminate (in the sense of terminating action) the output state is present in each action.

On the other hand, we can choose messages for threading actions. In this perspective, the agent's messages form the *agent's process*. The messages between actions define the *succession relation* between actions. The process runs along its messages while visited servers' states are the communication means with other agents. Because there can exist actions without an output message—the agent process can terminate.

$$\begin{aligned}
&B(s) = \langle \lambda_1, \lambda_2, \ldots \rangle \big| \forall_{i=1,\ldots} \lambda_i \in \Lambda; \\
&\lambda_1 = ((m,p),(m_1,p_1)) \vee \lambda_1 = ((m,p),(p_1)), p \in P_{ini} \cap P(s); \\
&\forall_{i>1} \lambda_{i-1} = ((m,p),(m_1,p_1)) \vee \lambda_{i-1} = ((m,p),(p_1)), \\
&\lambda_i = ((m_2,p_2),(m_3,p_3)) \vee \lambda_i = ((m_2,p_2),(p_3)), \\
&p_1,p_2 \in P(s), p_1 = p_2 - \text{the server process of the server } s\text{: the first action of the process} \\
&\text{has the initial state of the server } s \text{ on input, all the threading states appoint the server } s. \\
&C(a) = \langle \lambda_1, \lambda_2, \ldots \rangle \big| \forall_{i=1,\ldots} \lambda_i \in \Lambda; \\
&\lambda_1 = ((m,p),(m_1,p_1)) \vee \lambda_1 = ((m,p),(p_1)), m \in M_{ini} \cap M(a); \\
&\forall_{i>1} \lambda_{i-1} = ((m,p),(m_1,p_1)), \lambda_i = ((m_2,p_2),(m_3,p_3)) \vee \\
&\lambda_i = ((m_2,p_2),(p_3)), m_1, m_2 \in M(a), \\
&m_1 = m_2 - \text{the agent process of the agent } a\text{: the first action of the process has the initial} \\
&\text{message of the agent } a \text{ on input, all the threading messages appoint the agent } a.
\end{aligned} \tag{3.12}$$

We call the processes *asynchronous sequential*, because:

- action sequences come from their definition;
- asynchrony comes from the feature that always a state waits for a message or a message waits for a matching state (and, at the same time, a state waits for a matching message).

Note that:

- A process can have more than one starting node if more than one action can appear first in the process. For example, the process *Sprod* Fig. 3.2 can begin from the putting action (*p_doSth, p_neutral*)Λ(*put, produce*), of from doing-something-else action (*p_doSth, p_neutral*)Λ(*p_doSth, p_neutral*) which is a self-transition not shown in Fig. 3.2a (but shown in Figs. 3.1 and 3.2b).
- The process may be a branched graph if there is non-determinism in the action set: it may be possible to perform multiple actions for a given state or for a given message.

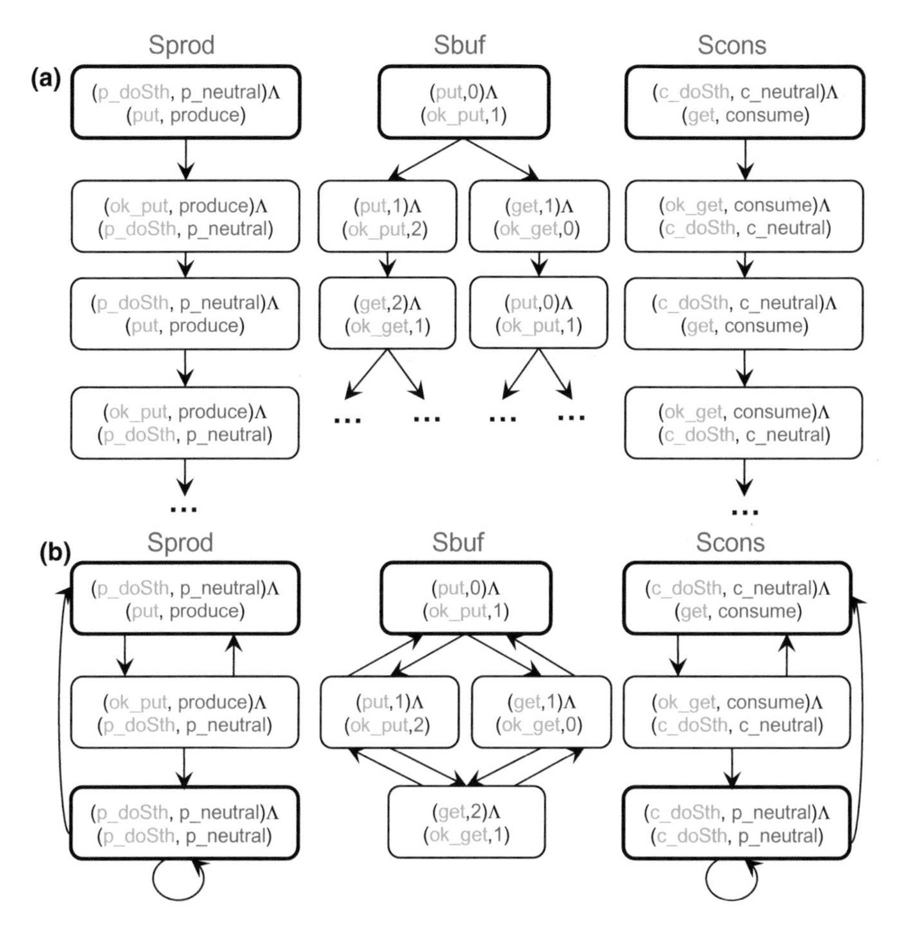

Fig. 3.2 **a** Fragments of sequences of actions in the server processes in the "Bounded buffer" system. Subsequences of actions (*p_doStr, p_neutral*)Λ(*p_doStr, p_neutral*) and actions (*c_doStr, c_neutral*)Λ(*c_doStr, c_neutral*) are suppressed, **b** succession relation

- Some paths in the graph may be finite if there is an action that terminates the agent's operation, or if no action is defined for a given state/message (this is not considered a process termination).

For the "Bounded buffer" example in Fig. 1.2, Fig. 3.2 presents server processes. The graph in Fig. 3.2b is converted by joining equal actions. Figure 3.3 shows one of the agent processes (*Aprod*).

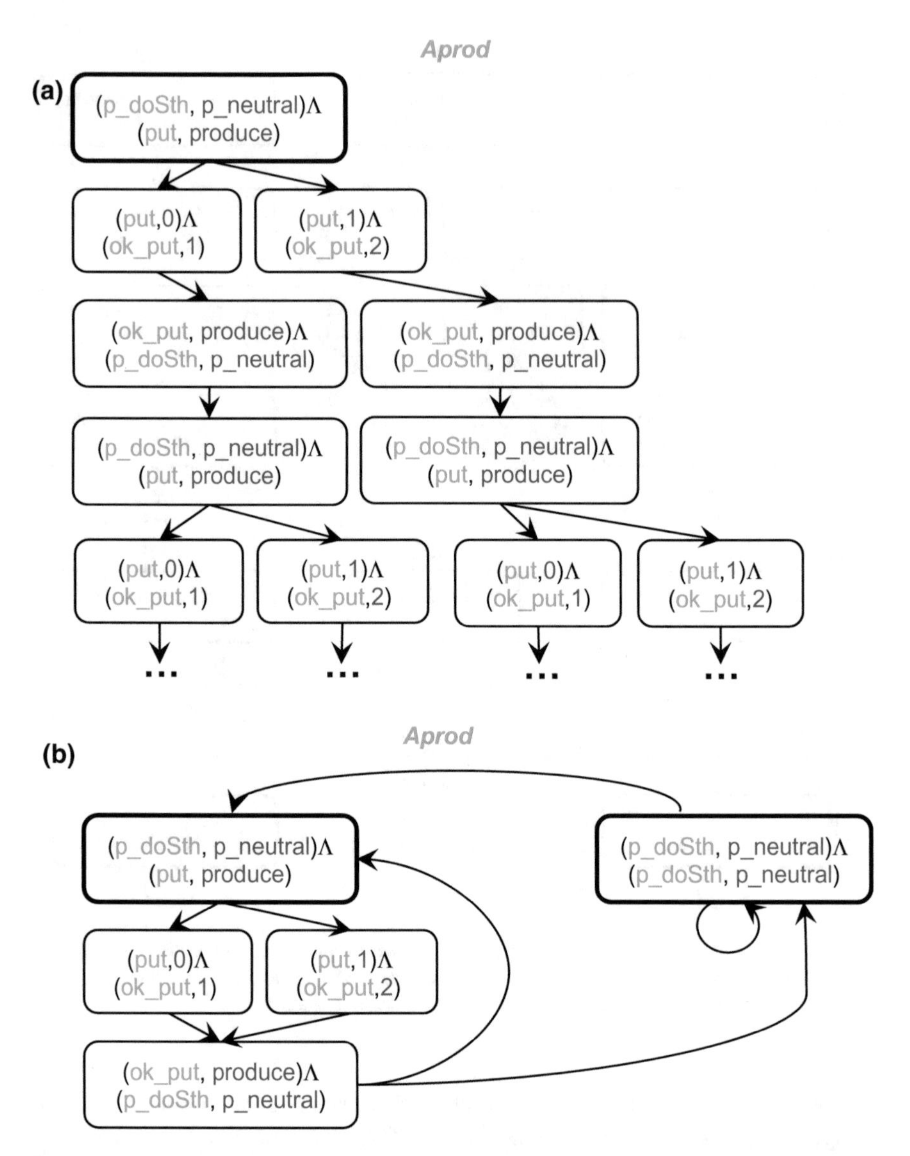

Fig. 3.3 **a** Fragment of graph of actions in the agent process "*Aprod*" in the "Bounded buffer" system. Subsequences of actions (*p_doStr, p_neutral*)*Λ*(*p_doStr, p_neutral*) are suppressed, **b** succession relation

3.4 Views of a Distributed System

The action sets of individual server processes are disjoint. The same concerns actions of agent processes. But there can be "orphan" actions, which do not belong to any server/agent. This can occur when an action has a state/message on input, which is neither initial nor present on output of any action belonging to any process. An orphan action denotes existence of unreachable, or "dead" code. To avoid such situations, in which there are actions not belonging to any process, other definition of processes is possible, grouping actions of individual servers/agents to sets:

$$\begin{aligned} B(s) &= \{\lambda \in \Lambda | \lambda = ((p,m),(p',m')) \vee \\ &\quad \lambda = ((p,m),(p')), p \in P(s)\} \\ &\quad -\text{the server process of the server } s \\ C(a) &= \{\lambda \in \Lambda | \lambda = ((p,m),(p',m')) \vee \\ &\quad \lambda = ((p,m),(p')), m \in M(a)\} \\ &\quad -\text{the agent process of the agent } a \end{aligned} \tag{3.13}$$

In the "Bounded buffer" example from Fig. 1.2, the processes defined as sets of actions are:

B(Sprod) = {(*p_doSth, p_neutral*)Λ(*p_doSth, p_neutral*), (*p_doSth, p_neutral*)Λ (*put, produce*), (*ok_put, produce*)Λ(*p_doSth, p_neutral*) }
B(Scons) = {(*c_doSth, c_neutral*)Λ(*p_doSth, p_neutral*), (*c_doSth, c_neutral*)Λ (*get, consume*), (*ok_get, consume*)Λ(*c_doSth, c_neutral*) }
B(Sbuf) = {(*put, 0*)Λ(*ok_put, 1*), (*put, 1*)Λ(*ok_put,2*), (*get, 1*)Λ(*ok_get, 0*), (*get, 2*)Λ(*ok_get, 1*) }
C(Aprod) = {(*p_doSth, p_neutral*)Λ(*p_doSth, p_neutral*), (*p_doSth, p_neutral*)Λ (*put, produce*), (*put, 0*)Λ(*ok_put, 1*), (*put, 1*)Λ(*ok_put,2*), (*ok_put, produce*)Λ (*p_doSth, p_neutral*) }
C(Acons) = {(*c_doSth, c_neutral*)Λ(*c_doSth, c_neutral*), (*c_doSth, c_neutral*)Λ (*get, consume*), (*get, 1*)Λ(*ok_get, 0*), (*get, 2*)Λ(*ok_get, 1*), (*ok_get, consume*)Λ (*c_doSth, c_neutral*) }.

Having all actions included in processes, we obtain the system decomposed into processes. Two decompositions into asynchronous sequential processes give the two system views:

$$\begin{aligned} \mathbf{B} &= \{B(s) | s \in S\} - \text{the server view} \\ \mathbf{C} &= \{C(a) | a \in A\} - \text{the agent view} \end{aligned} \tag{3.14}$$

In the example:

B = {*B(Sprod)*, *B(Scons)*, *B(Sbuf)*}
C = {*C(Aprod)*, *C(Acons)*}.

Many system properties can be verified using the model checking technique. This monograph focuses on deadlock and distributed termination properties (Chandy and Lamport 1985). Different features can be observed in different views (i.e., server view or agent view), as shown in Sect. 3.7.

3.5 IMDS as a Programming Language

In the basic IMDS definition, states and messages were defined as simple, indivisible entities. In the Dedan program designed for verification of distributed systems, IMDS is used as the specification language. To be able to define server types and agent types, and declare variables of these types, it is better to specify a server state as a pair (*server*, *value*). A message can also be defined as a pair (*agent*, *target server*). However, the agent can call various server operations, for example *P* and *V* operations on the semaphore. These operations will be called "services". Messages are therefore defined as triples (*agent*, *server*, *service*). In such a formulation, a message is an invocation of a server's service by an agent. Below is a formal definition of actions in terms of states as pairs and messages as triples.

The basic sets are servers, agents, values and services:

$S = \{s_1, s_2, \ldots\}$—the finite set of *servers*
$A = \{a_1, a_2, \ldots\}$—the finite set of *agents*
$V = \{v_1, v_2, \ldots\}$—the finite set of *values*
$R = \{r_1, r_2, \ldots\}$—the finite set of *services*.

The servers' states are defined as pairs (*server, value*) and the messages as triples (*agent, server, service*). In such a formulation, a message is an invocation of a server's service by an agent. An action is an execution of a service on a server in a context of an agent.

$P \subset S \times V$—the set of *states*
$M \subset A \times S \times R$—the set of *messages*.

The constraints on states and messages expressed in the new formulation in the input and output items of the action are:

$$\Lambda \subset (M \times P) \times (M \times P) \cup (M \times P) \times (P) | (m,p)\Lambda(m',p') \vee (m,p)\Lambda(p'),$$

$$m = (a, s, r) \in M,\, p = (s_1, v_1) \in P,$$
$$m' = (a_2, s_2, r_2) \in M,\, p' = (s_3, v_3) \in P, s_1 = s, s_3 = s, a_2 = a$$

The new definition of server and agent processes follows (equivalent to the previous one):

$B(s) = \{\lambda \in \Lambda \mid \lambda = (((a,s,r),(s,v)),\ ((a,s',r'),(s,v'))) \vee \lambda = (((a,s,r),(s,v)),\ ((s,v'))),$ $s' \in S,\ a \in A,\ v,v' \in V,\ r,r' \in R\ \}$—the server process of the server $s \in S$
$C(a) = \{\lambda \in \Lambda \mid \lambda = (((a,s,r),(s,v)),\ ((a,s',r'),(s,v'))) \vee \lambda = (((a,s,r),(s,v)),\ ((s,v'))),$ $s,s' \in S,\ v,v' \in V,\ r,r' \in R\ \}$—the agent process of the agent $a \in A$.

In the Dedan program, and in the rest of the monograph, we use the notation `s.v` and `a.s.r` for pairs (s,v) and triples (a,s,r), and the notation `{a.s.r, s.v}` → `{a,s'.r', s.v'}` for actions $((a,s,r),(s,v))\Lambda((a,s',r'),(s,v'))$. The

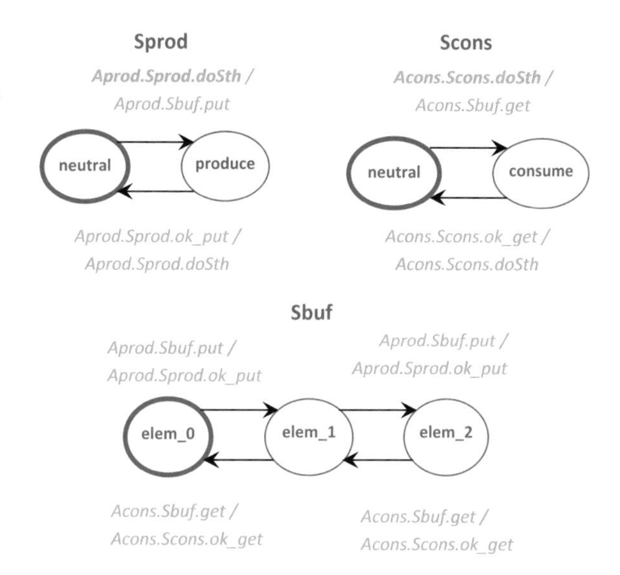

Fig. 3.4 "Bounded buffer" example: automata for buffer, producer and consumer (input and output symbols made of IMDS messages)

"Bounded buffer" example from Fig. 1.2, depicted graphically using the presented notation, is shown in Fig. 3.4.

The source code of the example is given below. It is a development of a system shown in Fig. 1.2. The example is presented informally in Fig. 3.4. It is the server view of a system consisting of a producer, a consumer and a buffer. Server names are preceded by '`S`', and agent names are preceded by '`A`'. Messages are input and output symbols on transitions leading from one server state to another. The rules for this figure (and consecutive ones) are[4]:

- every server is depicted as separate automaton-like graph of Mealy type (Dick and Yao 2014),
- nodes of an automaton are the server's states,
- initial node (initial state) is surrounded by bold ellipse,
- initial messages of the agents are in bold font,
- transitions are actions,
- transitions are labeled by input symbol (input message)/output symbol (output message).

In the example, three servers cooperate: the buffer containing 0, 1 or 2 elements, the producer spontaneously making new elements and then sending them to the buffer, and the consumer retrieving the elements from the buffer and destroying them. A message invoking a buffer's action is accepted if the operation is possible, for example putting to a full buffer is prohibited. Consider the upper-left action in `Sbuf` server: the buffer is empty (a value `elem_0` denotes 0 elements in the buffer)

[4]This form of graphical specification will be a basis for definition of Distributed Automata DA^3 in Chap. 8.

and the message of *Aprod* agent is pending at *Sbuf*. The message invokes a *put* service of *Sbuf* server. The execution of the action inserts an element to the buffer: it changes its state from a value *elem_0* to a value *elem_1*. On finishing of the execution of the *put* service, the *Sbuf* server issues a message of the same agent *Aprod* to the server *Sprod*, invoking its service *ok_put*. Note that in the state *elem_2* of the server *Sbuf* a message invoking a service *put* cannot be accepted. If it is received, it remains pending (it is not shown in the figure, see the definition of distributed server automata is Sect. 8.1).

The full source code of the buffer example, extended to sets of producers and consumers (instead of single producer and single consumer in Fig. 3.4) in the server view is as follows:

```
1   system   BUF_server_view;

2   #DEFINE N 2
3   #DEFINE M 2
4   #DEFINE K 2

5   server:   Sbuf(agents Aprod[N],Acons[M]; servers Sprod[N],Scons[M]),
6   services {put, get},
7   states   {elem0,elem[K]},
8   actions{
9   <i=1..N>       {Aprod[i].Sbuf.put, Sbuf.elem0} ->
                   {Aprod[i].Sprod[i].ok_put, Sbuf.elem[1]},
10   <i=1..N><j=1..K-1> {Aprod[i].Sbuf.put, Sbuf.elem[j]} ->
                        {Aprod[i].Sprod[i].ok_put, Sbuf.elem[j+1]},
11   <i=1..M><j=2..K>   {Acons[i].Sbuf.get, Sbuf.elem[j]} ->
                        {Acons[i].Scons[i].ok_get, Sbuf.elem[j-1]},
12   <i=1..M>           {Acons[i].Sbuf.get, Sbuf.elem[1]} ->
                          {Acons[i].Scons[i].ok_get, Sbuf.elem0}
13   }

14   server:    Sprod(agents Aprod; servers Sbuf),
15   services   {doSth,ok_put}
16   states     {neutral,prod}
17   actions    {
18   {Aprod.Sprod.doSth, Sprod.neutral} -> {Aprod.Sbuf.put, Sprod. prod}
19   {Aprod.Sprod.ok_put, Sprod.prod} -> {Aprod.Sprod.doSth, Sprod. neutral}
20   }

21   server:   Scons(agents Acons; servers Sbuf),
22   services  {doSth,ok_get}
23   states    {neutral,cons}
24   actions   {
25   {Acons.Scons.doSth, Scons.neutral} -> {Acons.Sbuf.get,Scons.cons}
26   {Acons.Scons.ok_get, Scons.cons} -> {Acons.Scons.doSth, Scons.neutral}
27   }

28   servers  Sbuf,Sprod[N],Scons[M];
29   agents   Aprod[N],Acons[M];

30   init ->   {
31   <j=1..N> Sprod[j](Aprod[j],Sbuf).neutral,
32   <j=1..M> Scons[j](Acons[j],Sbuf).neutral,
33            Sbuf(Aprod[1..N],Acons[1..M],Sprod[1..N],Scons[1..M]). elem0,

34   <j=1..N> Aprod[j].Sprod[j].doSth,
35   <j=1..M> Acons[j].Scons[j].doSth,
36   }.
```

The notation is intuitive; the (optional) header contains the system name (line 1). The server types are defined first (l.5,14,21), then servers (l.28), agents (l.29) and initialization part (l.30). A server type consists of formal parameters defining servers and agents used (l.14), server's services (l.15), states (l.16) and the actions of the server (l.9–12). Since the vectors in the Dedan program are numbered from 1, the 0th state of the buffer is defined separately. In initialization part, actual parameters (agents and servers) are passed to every server (l.31–33) and initial state of every server is defined after the dot. For every agent, an initial message is defined (l.34,35).

The system converted to the agent view is given below. It is a manual conversion, to preserve the constants *M*, *N* and *K*. This is the only difference between the manual and the automatic conversion: the program Dedan performs the conversion automatically, but the constants disappear in the process due to preprocessing.

```
1   system BUF_agent_view;

2   #DEFINE N 2
3   #DEFINE M 2
4   #DEFINE K 2
5   server:      Sbuf,
6   services     {put, get}
7   states       {elem0, elem[K]}
8   ;
9   server:      Sprod,
10  services     {doSth, ok_put}
11  states       {neutral, prod}
12  ;
13  server:      Scons,
14  services     {doSth, ok_get}
15  states       {neutral, cons}
16  ;
17  agent:       Aprod (servers Sbuf,Sprod),
18  actions      {
19  <j=1..K-1> {Aprod.Sbuf.put, Sbuf.elem[j]} -> {Aprod.Sprod. ok_put, Sbuf.elem[j+1]},
20               {Aprod.Sbuf.Sput, Sbuf.elem0} -> {Aprod.Sprod. ok_put, Sbuf.elem[1]},
21               {Aprod.Sprod.doSth, Sprod.neutral} -> {Aprod.Sbuf. put, Sprod.prod},
22             {Aprod.Sprod.ok_put, Sprod.prod} -> {Aprod.Sprod. doSth, Sprod.neutral},
23  };
24  agent:       Acons (servers buf,Scons),
25  actions      {
26         {Acons.Sbuf.get, Sbuf.elem[1]} -> {Acons.Scons. ok_get, Sbuf.elem0},
27  <j=2..K>     {Acons.Sbuf.get, Sbuf.elem[j]} -> {Acons.Scons. ok_get, Sbuf.elem[j-1]},
28               {Acons.Scons.ok_get, Scons.cons} -> {Acons.Scons. doSth, Scons.neutral},
29               {Acons.Scons.doSth, Scons.neutral} -> {Acons.Sbuf. get, Scons.cons},
30  };
31  agents       Aprod[M], Acons[N];
32  servers      Sbuf, Sprod[M], Scons[N];

33  init ->      {
34  <j=1..M>     Aprod[j](Sbuf,Sprod[j]).Sprod[j].doSth,
35  <j=1..N>     Acons[j](Sbuf,Scons[j]).Scons[1].doSth,
36               Sbuf.elem0,
37  <j=1..M>     Sprod[j].neutral,
38  <j=1..N>     Scons[j].neutral,
39  }.
```

The (optional) header contains the system name (line 1). The server types are defined first (l.5,9,13). They contain only service and state definitions, as the actions are attributed to the agents. The agent types are defined separately (l.17,24). Then, there are declarations of agents (l.31), servers (l.32) and initialization part (l.33). The agent type consists of formal parameters defining servers used (l.17) and the actions of the agent (l.19–22). In the initialization part, actual parameters (servers) are passed to every agent (l.34–35) and the initial message of every agent is defined. For every server, the initial state is defined (l.36–38). The formal syntax of IMDS is given in Appendix C.

The presented input of the Dedan program follows IMDS formulation and it is uncomfortable for programmers. Therefore, the Rybu preprocessor for imperative-like programming was developed by the students of ICS, WUT under the supervision of the author (Daszczuk et al. 2017). The students of second-year studies use the Rybu language in verification of their synchronization solutions. More details on Rybu are given in Sect. 5.8.

3.6 Semantics

The semantics of modeled system is defined by a Labeled Transition System (LTS, Sect. 3.2, (Reniers and Willemse 2011)), in which the nodes are system configurations and transitions are actions.

If the items p and m match in the configuration, then an action having p and m on input can be executed. We say that the action is *prepared* (or *enabled*) in the configuration containing p and m. The execution of the action is called *firing* the action. We assume interleaving semantics (Glabbeek and Goltz 1990; Winskel and Nielsen 1995), i.e., only one of the prepared actions is fired at a time.

Note that the actions prepared on the same server are always in conflict, as each of them “consumes” the current server’s state.

On the other hand, the actions prepared on separate servers are always independent. The firing of the action on a given server does not affect the prepared actions on other servers. They remain prepared, although the action can add a new prepared action on some server.

The Labeled Transition System ($LTS = <N, n_0, W>$, see Sect. 3.2) is defined by the set of *nodes* N, *root node* n_0 and the set of *transitions* W. The nodes are the system configurations (with the root node being the initial configuration) and the transitions are actions. Such an *LTS* contains all possible executions of the system.

Nondeterminism is modeled as diverging branches in the LTS. IMDS provides three kinds of nondeterminism:

- *nondeterminism between servers*—if there are prepared actions on distinct servers, the action to fire is chosen in nondeterministic way,
- *nondeterminism in server*—if the current state matches pending messages of more than one agent—an action to fire is chosen nondeterministically,
- *nondeterminism in agent*—if more than one action is prepared with matching state and message—an action to fire is chosen nondeterministically.

The system is assumed to act fair, i.e., if the action is prepared infinite number of times—it must be fired. In other words, if there is an escape from the loop in the LTS, then after the finite number of cycles in the loop, the escape action must be selected. Thus, no part of LTS is left as "dead code" because on unfairness.

Usually, the LTS is large, therefore it cannot be observed manually in its entirety. A tiny example "*different strokes—different folks*" is prepared to show the entire LTS. The system consists of three servers. There is a single *TV* set, and there are two people: the *boy* switches to *sport* while the *girl* switches to the *music* program. The *TV* server has two states: *sport* and *music*. The services of the *TV* server are pushing the buttons: *b1* and *b2* switching to *sport* and *music*, respectively. An initial state of *TV* should be chosen, say: *sport*. The server view is as follows (graphical representation in Fig. 3.5, the rules are the same as for Fig. 3.4):

```
System    strokes;

server:    TV (agents Aboy,Agirl; servers Sboy,Sgirl),
states     {sport, music}
services   {b1,b2}
actions    {
          {Aboy.TV.b1, TV.sport} -> {Aboy.Sboy.ok, TV.sport},
          {Aboy.TV.b1, TV.music} -> {Aboy.Sboy.ok, TV.sport},
          {Agirl.TV.b2, TV.sport} -> {Agirl.Sgirl.ok, TV.music},
          {Agirl.TV.b2, TV.music} -> {Agirl.Sgirl.ok, TV.music},
};
server:    Sboy (agents Aboy; servers TV),
states     {wait, watch}
services   {switch, ok}
actions    {
          {Agirl.Sgirl.switch, Sgirl.watch} -> {Agirl.TV.b2, Sgirl. wait},
          {Agirl.Sgirl.ok, Sgirl.wait} -> {Agirl.Sgirl. switch, Sgirl.watch},
};
server:    Sgirl (agents Agirl; servers TV),
states     {wait, watch}
services   {switch, ok}
actions    {
          {Aboy.Sboy.switch, Sboy.watch} -> {Aboy.TV.b1, Sboy.wait},
          {Aboy.Sboy.ok, Sboy.wait} -> {Aboy.Sboy.switch, Sboy. watch},
```

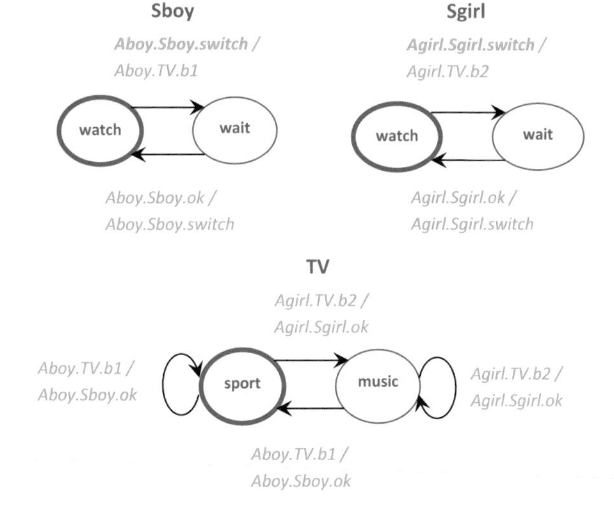

Fig. 3.5 "Strokes" example: automata for *Sboy*, *Sgirl* and *TV*

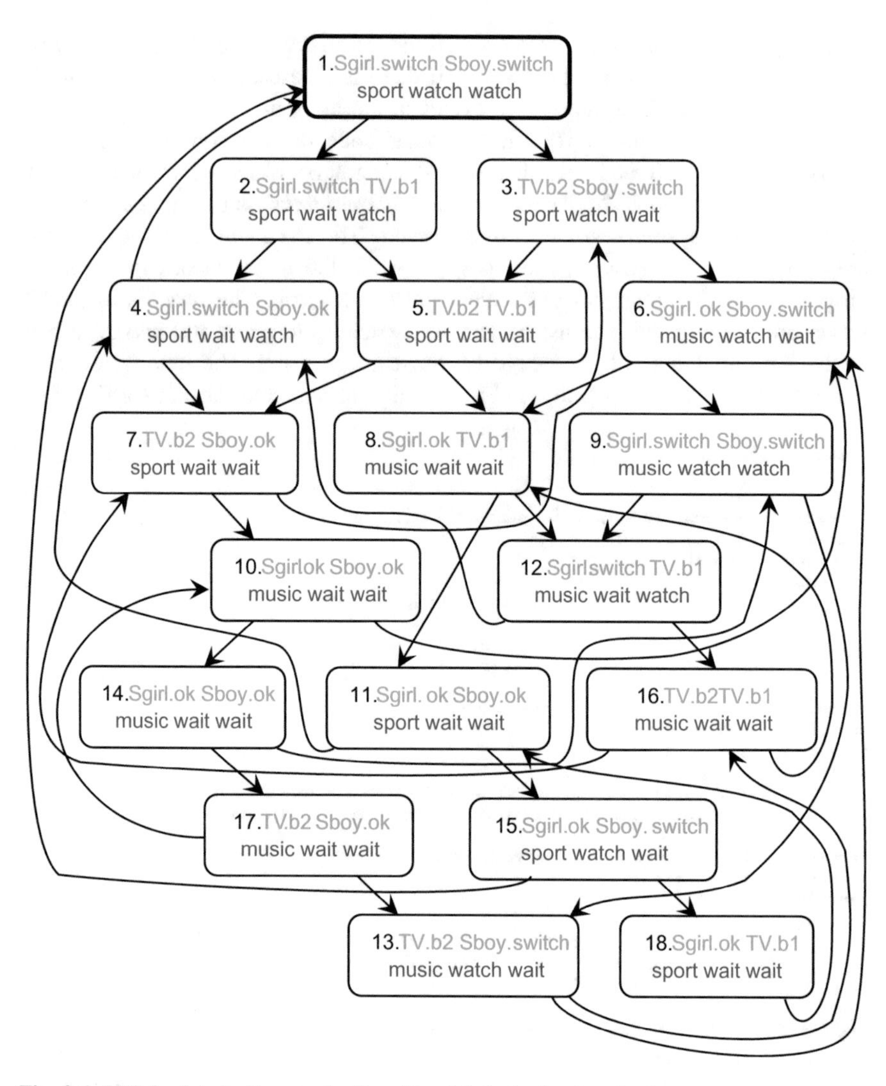

Fig. 3.6 LTS for "strokes" example. Transition labels (actions) omitted

```
};
servers  TV,Sboy,Sgirl;
agents   Agirl,Aboy;
init ->   {
         TV(Aboy,Agirl,Sboy,Sgirl).sport,
         Sboy(Aboy,TV).watch,
         Sgirl(Agirl,TV).watch,
         Aboy.Sboy.switch,
         Agirl.Sgirl.switch,
}.
```

The LTS of the *strokes* system is presented in Fig. 3.6. The nodes show configurations: pending messages of both agents (*Agirl* and *Aboy*) and states of all three servers (`TV`, `Sgirl` and `Sboy`). The initial configuration has a bold border. Every configuration has two successors, because of nondeterministic choice: "between servers" when agents have pending messages at different servers or "in server" when both agents have pending messages at `TV`. Nondeterminism in agent is not modeled in this example, but it can be observed in "philosophers" systems (Sect. 5.2), the philosopher when he thinks, may decide to continue thinking or taking a fork.

3.7 Deadlock and Termination

We define deadlock as a situation in which a process that is not terminated, cannot continue (at the moment and in the future). The deadlock should be defined separately for the server process and for the agent process, since they have different characteristics.

First consider a server process. The following situations can occur:

- the current state matches some of messages pending at the server—an action is prepared and it can be fired; the server process *runs*;
- the current state does not match any message pending at the server (or no message is pending)—a matching message may occur at the server in the future; the server process *waits*;
- neither a message is pending at the server nor a message will occur at the server in the future; the server process is *idle*;
- there are messages pending at the server, but the current state does not match any of the pending messages, and no matching message may occur at the server in the future; the server process is in *deadlock*.

The former two cases are normal operation of the server process. The third case means that the server will not change its state—the current state will be current forever. This is not an error, just no other server wants any service offered by the given server. The last case is a deadlock—some messages wait to be served, but neither the current state matches any of them nor it will match in the future. The cases can be distinguished during the analysis of the LTS.

The server process never terminates—the next state is obligatory in every action, although one may treat the idleness situation as a termination.

Consider the agent process. The cases are:

- the agent's message is pending at a server and it matches the current state of this server—the action is prepared and it can be fired; the agent *runs*;
- the agent's message is pending at a server and it does not match the current state of this server, but a matching state may occur in the future; the agent *waits*;
- the agent's message is pending at a server and neither the current state of this server matches the message nor such a state may occur in the future; the agent fell into a *deadlock*;

- there is no agent's message—the agent process is *terminated.*

The former two cases are the normal operation of the agent process. The third case is a deadlock—there is a message to be serviced but this will never happen. The last case means that the process is terminated. Again, the cases can be distinguished by the analysis of the LTS.

References

Chandy, K. M., & Lamport, L. (1985). Distributed snapshots: Determining global states of distributed systems. *ACM Transactions on Computer Systems, 3*(1), 63–75. https://doi.org/10.1145/214451.214456.

Chrobot, S., & Daszczuk, W. B. (2006). Communication dualism in distributed systems with petri net interpretation. *Theoretical and Applied Informatics, 18*(4), 261–278. URL: https://taai.iitis.pl/taai/article/view/250/taai-vol.18-no.4-pp.261.

Daszczuk, W. B. (2008). *Deadlock and termination detection using IMDS formalism and model checking, Version 2, ICS WUT Research Report No.2/2008.* arXiv:1710.08842.

Daszczuk, W. B., Bielecki. M., & Michalski, J. (2017). Rybu: Imperative-style preprocessor for verification of distributed systems in the Dedan environment. In *KKIO'17—Software Engineering Conference, Rzeszów, Poland,* 14–16 September, 2017. Polish Information Processing Society. arXiv:1710.02722.

Dick, G., & Yao, X. (2014). Model representation and cooperative coevolution for finite-state machine evolution. In *2014 IEEE Congress on Evolutionary Computation (CEC), Beijing, China* (pp. 2700–2707), 6–11 July, 2014. New York, NY: IEEE. https://doi.org/10.1109/cec.2014.6900622.

Penczek, W., Szreter, M., & Kuiper, R. (2000). Improving partial order reductions for universal branching time properties. *Fundamenta Informaticae, 43*(1–4), 245–267. Url: https://content.iospress.com/articles/fundamenta-informaticae/fi43-1-4-13.

Reniers, M. A., & Willemse T. A. C. (2011). Folk theorems on the correspondence between state-based and event-based systems. In *37th Conference on Current Trends in Theory and Practice of Computer Science, Nový Smokovec, Slovakia* (pp. 494–505), 22–28 January, 2011. Berlin Heidelberg: Springer. https://doi.org/10.1007/978-3-642-18381-2_41.

van Glabbeek, R. J., & Goltz, U. (1990). Equivalences and refinement. In *LITP Spring School on Theoretical Computer Science La Roche Posay, France*, LNCS 469 (pp. 309–333), 23–27 April, 1990. Berlin Heidelberg: Springer. https://doi.org/10.1007/3-540-53479-2_13.

Winskel, G., & Nielsen, M. (1995). *Models for concurrency. Handbook of logic in computer science* (vol.4). S. Abramsky, D. M. Gabbay, & T. S. E. Maibaum (Eds.). Oxford, UK: Oxford University Press. ISBN:0-19-853780-8.

Chapter 4
Model Checking of IMDS Specifications in the Dedan Environment

The LTS of a system can be analyzed manually or using graph algorithms. However, the LTS can be interpreted as the Kripke Structure (Clarke et al. 1999): finite set of *nodes* (we do not call them states to avoid ambiguity)—configurations, *root node*—initial configuration, a total *transition relation*—actions, naturally assignable *labeling* (see later). This allows the application of model checking techniques to analyze the graph, especially to find deadlocks and check termination of processes.

A communication deadlock occurs when the current server's state does not match any message pending at the server, and no matching message may occur at the server in the future. In other words, some messages are pending but no action will be fired on the server. We label the LTS nodes using the Boolean formulas (only configurations with *true* labels are described, *false* is for all remaining configurations):

- D_s—*true* in all configurations where at least one message is pending at the server s,
- E_s—*true* in all configurations where at least one action is prepared at the server s.

The formulas checking if the server s falls into a communication deadlock or to an idle state are given in Table 4.1, features *dds*(s) and *idle*(s), respectively. Properties are given both in LTL and CTL temporal logic, as the Dedan program cooperates with three external model checkers: Spin [using LTL (Holzmann 1995, 1997)], NuSMV [using CTL (Cimatti et al. 2000)], and Uppaal [using CTL (Behrmann et al. 2006)].

A resource deadlock (or deadlock over resources) occurs when the agent process' message is pending at the server but it will never match any state of this server. We need the following labelling of the LTS (F_a is for finding agent termination):

© Springer Nature Switzerland AG 2020

W. B. Daszczuk, *Integrated Model of Distributed Systems*, Studies in Computational Intelligence 817, https://doi.org/10.1007/978-3-030-12835-7_4

Table 4.1 Temporal formulas for finding various situations in processes

Property	LTL	CTL
Communication deadlock in server *s: dds*(*s*)	$\Diamond\Box\,(D_s \wedge \neg E_s)$	**EF AG** $(D_s \wedge \neg E_s)$
Server *s* idle*: idle*(*s*)	$\Diamond\Box\,(\neg D_s)$	**AF AG** $(\neg D_s)$
Resource deadlock in agent *a: dda*(*a*)	$\Diamond\Box\,(D_a \wedge \neg E_a)$	**EF AG** $(D_a \wedge \neg E_a)$
Termination of agent *a: term*(*a*)	$\Diamond\,(F_a)$	**AF** (F_a)

- D_a—*true* in all configurations where a message of the agent *a* is pending,
- E_a—*true* in all configurations where an action is prepared with a message of the agent *a*,
- F_a—*true* in all configurations where a terminating action (with a message of the agent *a* on input) is prepared.

The formula that checks if the agent *a* falls into a resource deadlock is given in Table 4.1, feature *dda(a).*

An agent termination *term(a)* (no message of the agent pending) is also included. Of course, other properties can be added, provided that the proper labeling is applied on the LTS nodes.

Note that finding deadlocks concerns individual processes, not the entire system. This feature differs the presented technique from other deadlock detection approaches by static model checking.

The presented IMDS formalism, combined with model checking using general formulas, satisfies the desired features in modeling and verification of distributed systems:

- *Communication duality*: each system can be decomposed either to server processes communicating via messages (the action's output state is the server's process carrier, while the output message is a communication means between servers) or to agent processes communicating via servers' states (the action's output message is the agent's process carrier, while the output state is a communication means between the agents).
- *Locality*: each action on the server is executed on a basis of the current state of this server and the set of messages pending at this server. The server does not have access to any global or non-local variables. No external event (except messages received by the server) affects the behavior of the server.
- *Autonomy*: each server autonomously decides, which of the prepared actions will be executed and when (the LTS contains all possible scenarios). In other words, the servers autonomously decide if and when the communication would be accepted and what activities it will result.
- *Asynchrony*: the server accepts the message when it is ready for it, otherwise the message is pending; there is no synchrony in the model—no simultaneous activities of servers or agents, such as synchronous transitions on common symbols in Büchi automata (Holzmann 1995) or Timed Automata (Alur and Dill 1994), synchronization on *send* and *receive* operations in CSP (Hoare 1978;

Lanese and Montanari 2006), Occam (May 1983) or Uppaal timed automata (Behrmann et al. 2006), synchronous operations on complementary input and output *ports* in CCS (Milner 1984; Lanese and Montanari 2006). Autonomy of servers and agents is implemented by means of asynchronous operations: sending a message to the server is the only way to influence its behavior; setting a new value of the server's state is the only way to influence the behavior of agents visiting the server.

- *Asynchronous channels*: communication between the servers is unidirectional (communication in the opposite direction, if present, has its own channel), and may seem synchronous, because the message appears at a receiving server just when it is sent. However, asynchrony is modeled by possible deferring the reaction to the message (the message may wait for a long time before it is accepted). In the timed model (Chap. 10), the propagation time can be directly assigned to the communication channel.
- *Automated verification*: the temporal formulas are used to locate communication deadlocks in individual server processes, servers' idleness, resource deadlocks in agent processes and agents termination—regardless of a structure of the verified system—that is why they are the basis for building of automatic verifier Dedan. The program can be used without knowledge of temporal logics and model checking. Temporal logic formulas are "wired" in Dedan itself.

References

Alur, R., & Dill, D. L. (1994). A theory of timed automata. *Theoretical Computer Science, 126*(2), 183–235. https://doi.org/10.1016/0304-3975(94)90010-8.

Behrmann, G., David, A., & Larsen, K. G. (2006). *A Tutorial on Uppaal 4.0*. Aalborg, Denmark. url: http://www.it.uu.se/research/group/darts/papers/texts/new-tutorial.pdf.

Cimatti, A., Clarke, E., … Roveri, M. (2000). NUSMV: A new symbolic model checker. *International Journal on Software Tools for Technology Transfer*, *2*(4), 410–425. https://doi.org/10.1007/s100090050046.

Clarke, E. M., Grumberg, O., & Peled, D. (1999). *Model checking*. Cambridge, MA: MIT Press. ISBN: 0-262-03270-8.

Hoare, C. A. R. (1978). Communicating sequential processes. *Communications of the ACM, 21*(8), 666–677. https://doi.org/10.1145/359576.359585.

Holzmann, G. J. (1995). Tutorial: Proving properties of concurrent systems with SPIN. In *6th International Conference on Concurrency Theory, CONCUR'95, Philadelphia, PA, 21–24 Aug. 1995* (pp. 453–455). Heidelberg: Springer. https://doi.org/10.1007/3-540-60218-6_34.

Holzmann, G. J. (1997). The model checker SPIN. *IEEE Transactions on Software Engineering, 23*(5), 279–295. https://doi.org/10.1109/32.588521.

Lanese, I., & Montanari, U. (2006). Hoare vs Milner: Comparing synchronizations in a graphical framework with mobility. *Electronic Notes in Theoretical Computer Science, 154*(2), 55–72. https://doi.org/10.1016/j.entcs.2005.03.032.

May, D. (1983). OCCAM. *ACM SIGPLAN Notices, 18*(4), 69–79. https://doi.org/10.1145/948176.948183.

Milner, R. (1984). *A calculus of communicating systems*. Heidelberg: Springer. LNCS vol. 9, https://doi.org/10.1007/3-540-10235-3.

Chapter 5
Deadlock Detection Examples: The Dedan Environment at Work

5.1 "Two Semaphores" Model

The example of deadlock detection is presented for the system in which two distributed computations, each one running on its own server, use two semaphores. Each semaphore resides on a separate server.

```
A1:              A2:
sem1.wait;       sem2.wait;
sem2.wait;       sem1.wait;
sem1.signal;     sem2.signal;
sem2.signal;     sem1.signal;
stop             stop
```

This system falls into a total deadlock when *A1* holds *sem1* and waits for *sem2* and *A2* holds *sem2* and waits for *sem1*. But the situation changes if we add another agent process *A3* on its own server *r*, which simply loops and performs its own calculations. In the deadlock, processes *A1* and *A2* cannot continue, but *A3* still runs. We introduce calculations made by *A3* symbolically as an endless execution of `left` and `right` services on the server *r*:

```
A3:
loop {
      r.left;
      r.right
}
```

The system is not in a total deadlock, it is a partial deadlock, which is difficult to identify using model checking techniques (no system state is a "state with no

© Springer Nature Switzerland AG 2020

W. B. Daszczuk, *Integrated Model of Distributed Systems*, Studies in Computational Intelligence 817, https://doi.org/10.1007/978-3-030-12835-7_5

future"). In the system, the termination of the agents `A1` and `A2` (which are designed to terminate after some calculations) is not inevitable (the deadlock is not the termination). Using IMDS, partial deadlock of `A1` and `A2` can be identified using the proposed general formulas that show the lack of termination of these agents. The agent `A3` does not terminate as well—but this is expected because it works in an infinite loop in which there are no terminating actions.

The verification is carried out in the Dedan (Deadlock analyzer, ("Dedan," n.d.)) environment which consists of a specification part and TempoRG model checker (Daszczuk, 2003). The exemplary system consisting of two semaphores is presented graphically (informally) in Fig. 5.1a. The rules are the same as in the case of Fig. 3.4, except for the parameters showing the relationship between the specific instances of automata-like types. In Figs. 3.4 and 3.5 the servers are represented as

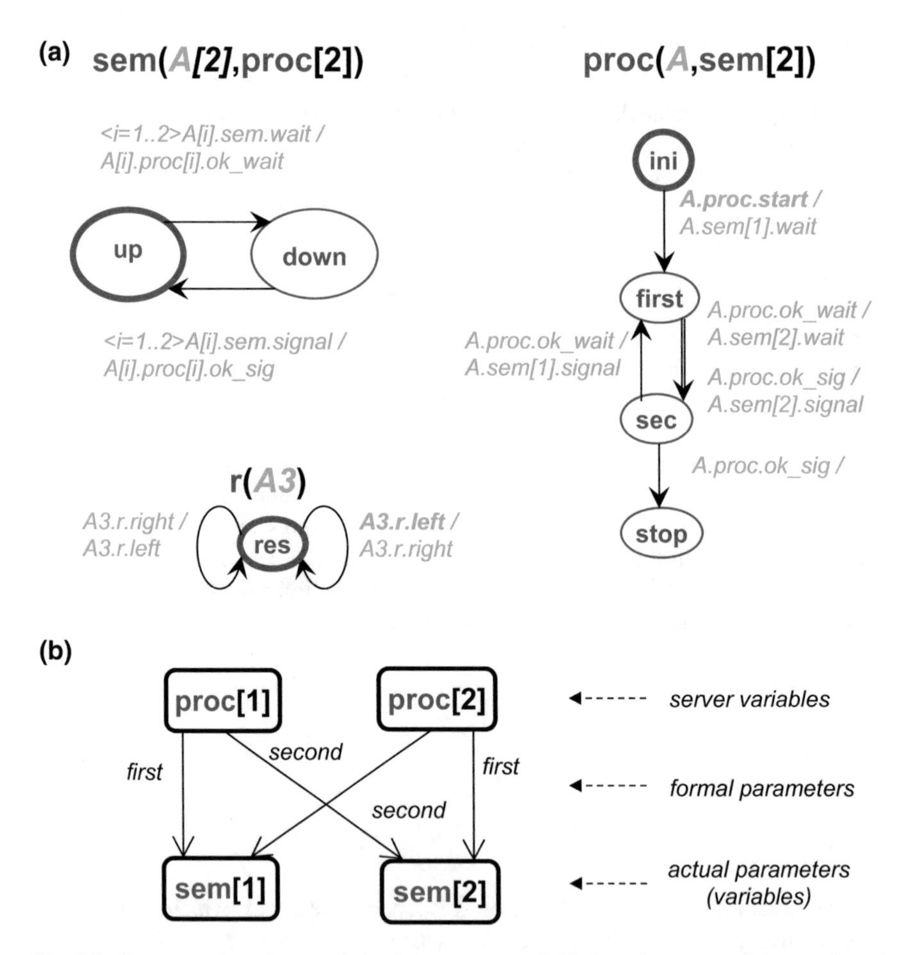

Fig. 5.1 "two semaphores" example: **a** Automata types (with formal parameters) **b** mapping of actual parameters (switched for *proc[2]*)

automata. In Fig. 5.1a, the server types are shown, therefore the automata names are supplemented by the sets of formal parameters. The system is presented below in IMDS notation, which is the input notation of Dedan. The specification is in the server view. It is simply grouping of the actions in individual servers. The system is defined as a sequence of server type specifications (enclosed by **`server`** ...`};` – lines 2-9, 10-19, 20-26*)*, server and agent instances (variables) declaration (***`agents`*** ...*,* ***`servers`*** ... – lines 27,28) and the initial configuration phrase (***`init`*** → *{...}* – lines 29-36). The server type heading contains a set of formal parameters: agents and servers used in the actions of the server type. Formal parameters can be vectors, as *`A[2]`* and *`proc[2]`* (l.2). Each server type contains the set of services (l.3,11,21), the set of states (l.4,12,22) and the set of assigned actions (in arbitrary order, l.6-8, 14-18, 24-25). The services and states can be vectors. To obtain a compact definition, repeaters can precede the actions in the server type definition (up to 3 for every action definition, l.6-8). The repeater is an integer variable with a defined range. The action definition is repeated for every value of a preceding repeater. If multiple repeaters are applied, the Cartesian product of their values is used. The indices of agents, states and services indicate individual instances of the vectors (l.6-8).

The server and agent variables can be organized in vectors (l.27,28). In the initialization part (l.29), the actual parameters are bound with formal parameters (again, repeaters and indices may be used—l.30-34). The initial server states and initial agent messages are also defined.

An important feature of the example is that actual parameters of *`proc`* servers are switched for *`proc2`*, therefore *`A[1]`* uses first semaphore, then second, and *`A[2]`* uses second semaphore, then first. This is graphically shown in Fig. 5.1b: the servers are represented by rounded boxes, and the arrows represent the parameter binding: labels of arrows show the order of the formal parameters and the arrowheads indicate the actual parameters bound with the formal parameters—switched for *`proc2`*.

```
 1. system   two_sem_server_view;
 2. server:  sem (agents A[2]; servers proc[2]),
 3. services {wait, signal},
 4. states   {up, down},
 5. actions  {
 6. <j=1..2> {A[j].sem.wait, sem.up} -> {A[j].proc[j].ok_wait, sem.down},
 7. <j=1..2> {A[j].sem.signal, sem.down} -> {A[j].proc[j].ok_sig, sem.up},
 8. <j=1..2> {A[j].sem.signal, sem.up} -> {A[j].proc[j].ok_sig, sem.up},
 9. };
10. server:  proc (agents A; servers sem[2]),
11. services {start, ok_wait, ok_sig},
12. states   {ini, first, sec, stop},
13. actions  {
14.          {A.proc.start, proc.ini} -> {A.sem[1].wait, proc.first},
15.          {A.proc.ok_wait, proc.first} -> {A.sem[2].wait, proc.sec},
16.          {A.proc.ok_wait, proc.sec} -> {A.sem[1].signal, proc.first},
17.          {A.proc.ok_sig, proc.first} -> {A.sem[2].signal, proc.sec},
18.          {A.proc.ok_sig, proc.sec} -> {proc.stop},
19. };

20. server:  r (agents A3),
21. services {left, right},
```

```
22. states    {res},
23. actions   {
24.           {A3.r.left, r.res} -> {A3.r.right, r.res},
25.           {A3.r.right, r.res} -> {A3.r.left, r.res},
26. };

27. agents:  A[2], A3;
28. servers:  sem[2], proc[2], r;

29. init ->   {
30. <j=1..2> A[j].proc[j].start,
31.           A3.r.left,
32.           proc[1](A[1],sem[1],sem[2]).ini;
33.           proc[2](A[2],sem[2],sem[1]).ini;
34. <j=1..2> sem[j](A[1],A[2],proc[1],proc[2]).up,
35.           r(A3).res,
36. }.
```

Note that the above system is defined in the server view, because the actions concerning a given server type are grouped inside its definition. The server states: *up, down, first, sec* are hidden inside the server types (not used outside the servers containing them), and the communication is performed via messages (*wait, ok_wait, ...*). In Dedan, the system can be automatically converted to the agent view. In the agent view, the messages are hidden inside the agent types, and agents communicate via servers' states. The states and services are the server attributes (lines 3, 4, 6, 7, 9, 10) just like in the server view. The actions are grouped in the agent types (l.13-21, 25-33, 37-38). The agent type *A* is split during the conversion into two types *A* and *A__1*. The reason is the switching of actual parameters (*sem[1]* and *sem[2]*) in the case of the second agent.

```
 1. system    two_sem_agent_view;

 2. server:    sem,
 3. services  {wait, signal},
 4. states     {up, down};

 5. server:    proc,
 6. services  {start, ok_wait, ok_sig},
 7. states     {ini, first, sec, stop};

 8. server:    r,
 9. services  {left, right},
10. states     {res};

11. agent:     A (servers proc,sem[2]),
12. actions    {
13.            {A.proc.start, proc.ini} -> {A.sem[1].wait, proc.first},
14.            {A.sem[1].wait, sem[1].up} -> {A.proc.ok_wait, sem[1].down},
15.            {A.proc.ok_wait, proc.first} -> {A.sem[2].wait, proc.sec},
16.            {A.sem[2].wait, sem[2].up} -> {A.proc.ok_wait, sem[2].down},
17.            {A.proc.ok_wait, proc.sec} -> {A.sem[1].signal, proc.first},
18.            {A.sem[1].signal, sem[1].down} -> {A.proc.ok_sig, sem[1].up},
19.            {A.proc.ok_sig, proc.first} -> {A.sem[2].signal, proc.sec},
20.            {A.sem[2].signal, sem[2].down} -> {A.proc.ok_sig, sem[2].up},
21.            {A.proc.ok_sig, proc.sec} -> {proc.stop},
22. };

23. agent:     A__1 (servers proc,sem[2]),
24. actions    {
25.           {A__1.proc.start, proc.ini} -> {A__1.sem[1].wait, proc.first},
26.           {A__1.sem[1].wait, sem[1].up} -> {A__1.proc.ok_wait, sem[1].down},
```

```
27.        {A__1.proc.ok_wait, proc.first} -> {A__1.sem[2].wait, proc.sec},
28.        {A__1.sem[2].wait, sem[2].up} -> {A__1.proc.ok_wait, sem[2].down},
29.        {A__1.proc.ok_wait, proc.sec} -> {A__1.sem[1].signal, proc.first},
30.        {A__1.sem[1].signal, sem[1].down} -> {A__1.proc.ok_sig, sem[1].up},
31.        {A__1.proc.ok_sig, proc.first} -> {A__1.sem[2].signal, proc.sec},
32.        {A__1.sem[2].signal, sem[2].down} -> {A__1.proc.ok_sig, sem[2].up},
33.        {A__1.proc.ok_sig, proc.sec} -> {proc.stop},
34. };

35. agent:   A3 (servers r),
36. actions  {
37.          {A3.r.left, r.res} -> {A3.r.right, r.res},
38.          {A3.r.right, r.res} -> {A3.r.left, r.res},
39. };

40. agents:  A, A__1, A3;
41. servers: sem[2], proc[2], r;

42. init ->   {
43. <j=1..2> proc[j].ini,
44. <j=1..2> sem[j].up,
45.          r.res,
46.         A(proc[1],sem[1..2]).proc[1].start,
47.         A__1(proc[2],sem[2],sem[1]).proc[2].start,
48.         A3(r).r.left,
49. }.
```

The Dedan program elaborates the global reachability space and launches TempoRG model checker to verify CTL formulas. Alternatively, the specification in input form of Spin, NuSMV or Uppaal is prepared for verification under the external model checker. The results of the verification are—communication deadlocks:

- server `sem[1]`—*true* (deadlock in `sem[1]`),
- server `sem[2]`—*true* (deadlock in `sem[2]`),
- server `proc[1]`—*false* (no deadlock in `proc[1]`),
- server `proc[2]`—*false* (no deadlock in `proc[2]`),
- server `r`—*false* (no deadlock in `r`),

resource deadlocks:

- agent `A` (`A[1]` in the server view)—*true* (deadlock in `A`),
- agent `A__1` (`A[2]` in the server view)—*true* (deadlock in `A__1`),
- agent `A3`—*false* (no deadlock in `A3`),

agent termination:

- agent `A` (`A[1]` in the server view)—*false* (`A` may not terminate),
- agent `A__1` (`A[2]` in the server view)—*false* (`A__1` may not terminate),
- agent `A3`—*false* (`A3` does not terminate, which is expected as it contains an endless loop).

To sum up, in our case the agent processes `A` and `A__1` (`A[1]` and `A[2]` in the server view) fall into a resource deadlock while the agent `A3` does not. The server processes `sem[1]` and `sem[2]` fall into a communication deadlock while the server `r` does not. No agent achieves an inevitable termination.

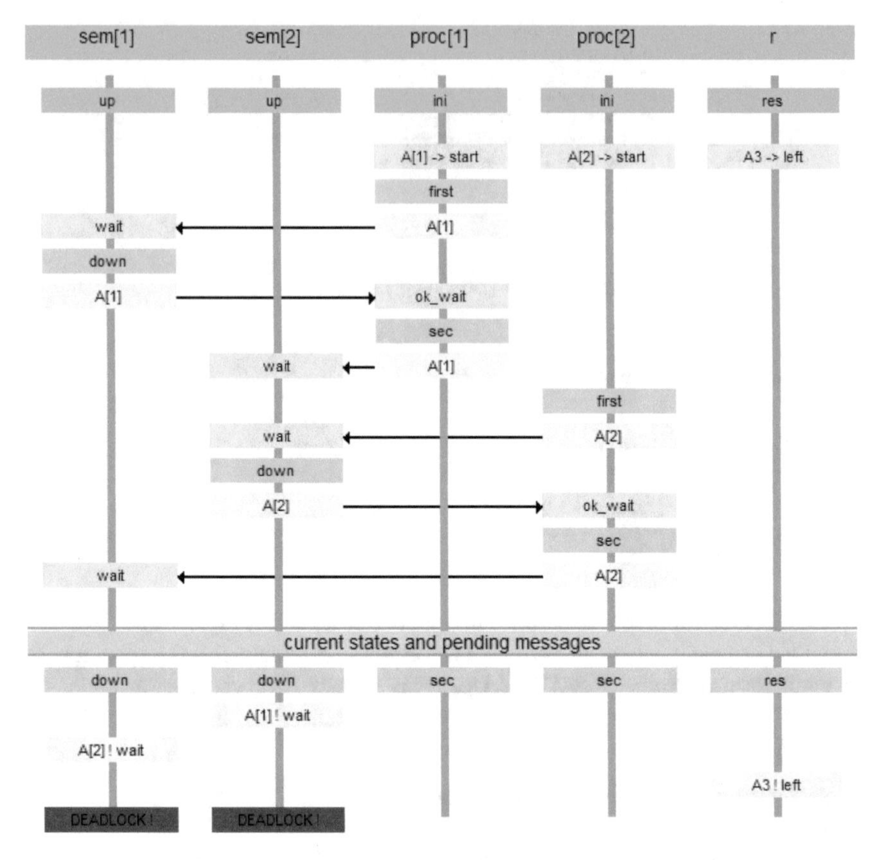

Fig. 5.2 Common communication deadlock in *sem[1]* and *sem[2]* in server view

The Dedan program generates the counterexamples (in the case of deadlock or lack of termination) or witnesses (in the case of successive termination) and displays them in readable format. The deadlock or termination can be checked for individual processes or for a given set of processes at the same time (common deadlock/termination). The example of the common communication deadlock of `sem[1]` and `sem[2]` in the server view is shown in Fig. 5.2. This is the output file from the Dedan program. The sequence diagram is redrawn in Fig. 5.3 for readability. Note the situation in which two subsequent messages invoke the `wait` service on the `sem[2]` server (surrounded by the circle). Both messages match the current state of `sem[2]`: `up`. The choice of an action to fire is nondeterministic. If the message originated from `A[1]` is selected—no deadlock occurs. However, model checking finds a counterexample that leads to the violation of the deadlock-safety rule, therefore the sequence is displayed in which the message of `A[2]` is selected.

Resource deadlock of the agent processes in the agent view is shown in Fig. 5.4. The lack of termination of all three agent processes is shown in Fig. 5.5. Recall that

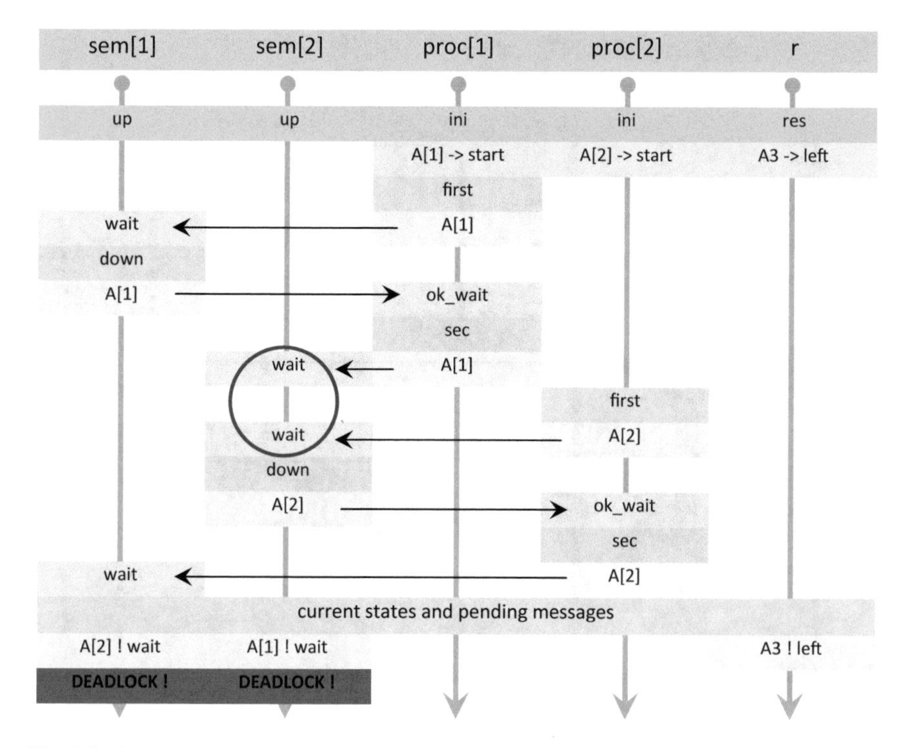

Fig. 5.3 Common communication deadlock in *sem[1]* and *sem[2]* in the server view. The circle surrounds two consecutive messages that are pending at the *sem[2]* server

agent type `A` in the server view is split to two agent types `A` and `A__1` when converted to the agent view. Agent variables have the same names as their types: `A` and `A__1`.

The counterexample or the witness is presented as a sequence diagram-like chart. The heading shows the names of servers on pink background (and agents on green background in the agent view). The upper part shows the initial states of all servers on pink background, and the sequences of states and messages leading to incorrect or desirable situation in a given process: deadlock, termination or lack of termination. A state is displayed simply as its identifier with a light blue background. The agent that causes the server process to enter the given state is shown on a dark blue background. A message is displayed with the agent identifier on light yellow background on send, and with service name on yellow background on receive. The middle part of the picture is a sequence that repeats in a loop infinitely (if there is one - in the example it is visiting the `left` and `right` states in the `A3` agent process). The last part shows the deadlock, termination or lack of termination configuration: all servers' states and all pending messages are displayed.

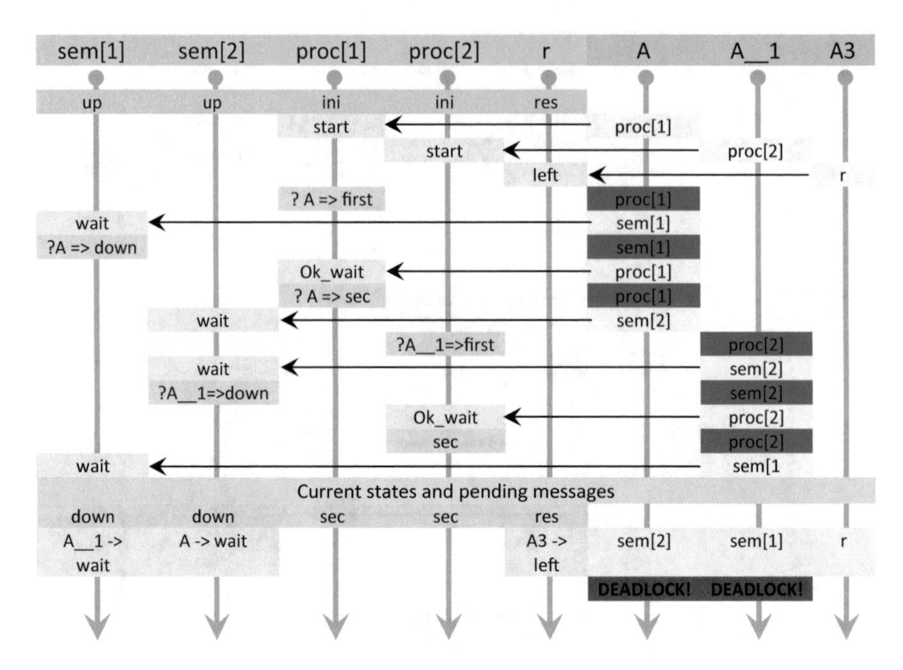

Fig. 5.4 Resource deadlock of agents in the agent view

5.2 Dijkstra's Philosophers

The problem of "5 philosophers" is a good example of the occurrence of deadlock (Dijkstra 1971). The example is shown in the distributed version, i.e., each philosopher is an agent sitting on his own chair, which is a separate server. Forks are the resources represented by separate servers. The structure of the agents and servers is shown in Fig. 5.6. The servers are depicted as rounded boxes and the agents are ovals. Message flow between servers is shown as arrows. The agents are connected to the servers on which they are started. The IMDS code for 5 philosophers (the server view) is:

```
1. system   phil;

2. server:  fork(agents ph[2]; servers chair[2]),
3. services {take_first, take_second, release_first, release_second},
4. states   {taken, free},
5. actions {
6. <j=1..2> {ph[j].fork.take_first, fork.free} ->
               {ph[j].chair[j].may_take_second, fork.taken},
7. <j=1..2> {ph[j].fork.take_second, fork.free} ->
               {ph[j].chair[j].may_eat, fork.taken},
8. <j=1..2> {ph[j].fork.release_first, fork.taken} ->
               {ph[j].chair[j].may_release_second, fork.free},
9. <j=1..2> {ph[j].fork.release_second, fork.taken} ->
               {ph[j].chair[j].may_think, fork.free},
10. };
```

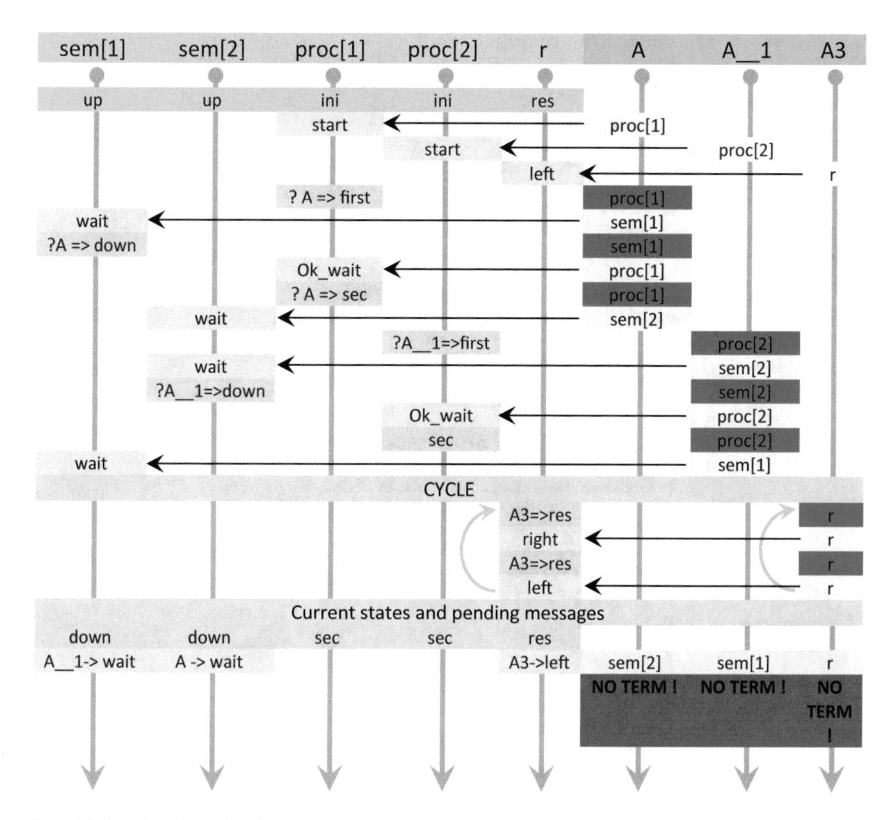

Fig. 5.5 Lack of termination of all three agents in the agent view

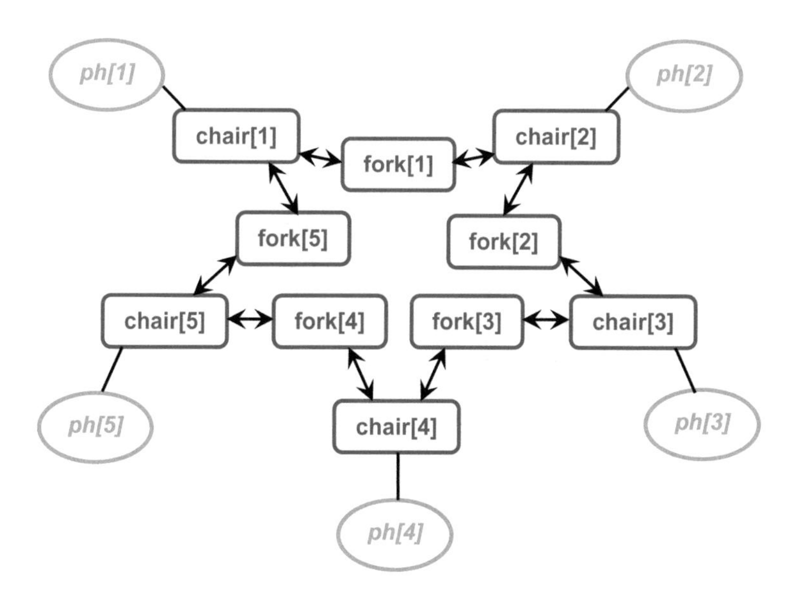

Fig. 5.6 Agent/server structure of "five philosophers" system

```
11. server:  chair(agents ph; servers fork[2]),
12. services {may_eat, may_think, may_take_second, may_release_second},
13. states   {think, waits_eat, eat, first_left, first_right},
14. actions  {
15.          {ph.chair.may_think, chair.think} ->
                {ph.chair.may_think, chair.think},
16.          {ph.chair.may_think, chair.think} ->
                {ph.fork[1].take_first, chair.first_left},
17.          {ph.chair.may_think, chair.think} ->
                {ph.fork[2].take_first, chair.first_right},
18.          {ph.chair.may_take_second, chair.first_left} ->
                {ph.fork[2].take_second, chair.waits_eat},
19.          {ph.chair.may_take_second, chair.first_right} ->
                {ph.fork[1].take_second, chair.waits_eat},
20.          {ph.chair.may_eat, chair.waits_eat} ->
                {ph.chair.may_eat, chair.eat},
21.          {ph.chair.may_eat, chair.eat} ->
                {ph.chair.may_eat, chair.eat},
22.          {ph.chair.may_eat, chair.eat} ->
                {ph.fork[1].release_first, chair.first_left},
23.          {ph.chair.may_eat, chair.eat} ->
                {ph.fork[2].release_first, chair.first_right},
24.          {ph.chair.may_release_second, chair.first_left} ->
                {ph.fork[2].release_second, chair.think},
25.          {ph.chair.may_release_second, chair.first_right} ->
                {ph.fork[1].release_second, chair.think},
26. };

27. servers: fork[5],chair[5];
28. agents: ph[5];

29. init -> {fork[1](ph[5,1],chair[5,1]).free,
30.           fork[2](ph[1,2],chair[1,2]).free,
31.           fork[3](ph[2,3],chair[2,3]).free,
32.           fork[4](ph[3,4],chair[3,4]).free,
33.           fork[5](ph[4,5],chair[4,5]).free,
34.           chair[1](ph[1],fork[1,2]).think,
35.           chair[2](ph[2],fork[2,3]).think,
36.           chair[3](ph[3],fork[3,4]).think,
37.           chair[4](ph[4],fork[4,5]).think,
38.           chair[5](ph[5],fork[5,1]).think,
39. <i=1..5>  ph[i].chair[i].may_think,
40. }.
```

Each philosopher *ph[i]* (line 28) sitting on a *chair* (type declared in l.11) *thinks* then *eats* (states declared in l.13). He needs two *forks* (parameters in l.11) to *eat*. Each *fork* (type in l.2) is a shared resource of two neighboring philosophers. More specifically, the philosopher *ph[i] thinks* for undefined time (l.15), then he decides nondeterministically to take *forks*, *first_left* (l.16) or *first_right* (l.17). After granting him the first *fork* (service *may_take_second*, l.12,18,19), the philosopher asks for the second one and waits for it in *waits_eat* state (l.13,18,19). Granting the second fork *may_eat* (l.20) causes the philosopher to *eat* (l.21), and then to release both *forks*, again in arbitrary order (l.22-25).

The automata-like model is presented in Fig. 5.7. The *forks* are *taken* in arbitrary order (*first_left* or *first_right*, nondeterministically, l.16,17), which causes a deadlock when all philosophers *take* their left *forks* first or when they take their right *forks* first. Double arrows show multiple actions between the states.

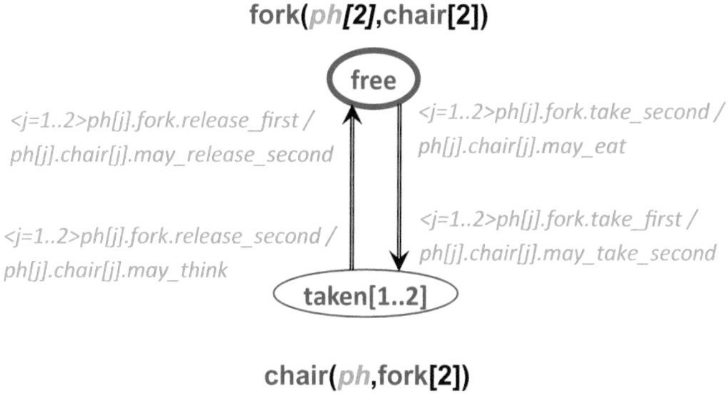

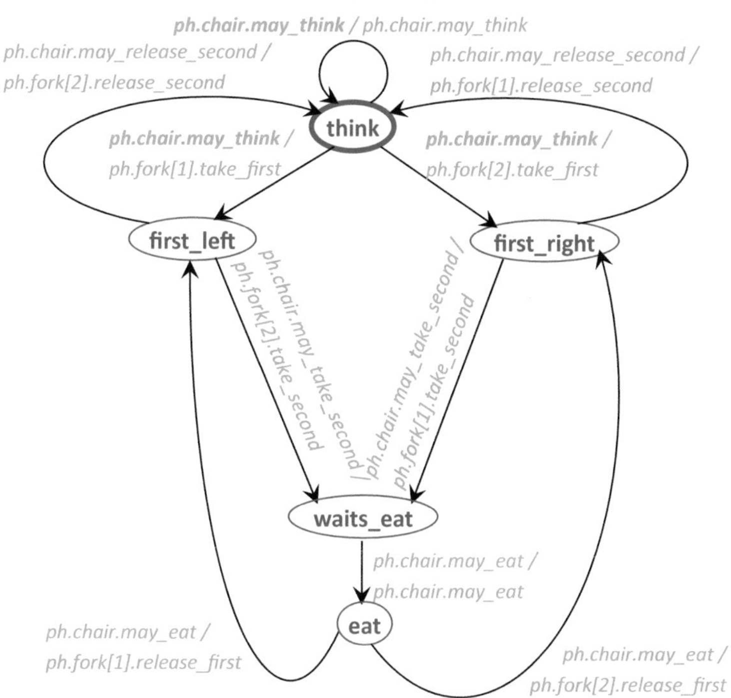

Fig. 5.7 Automata of *fork* and *chair* server types

The analysis of the system under Dedan shows a deadlock as expected.

The first solution to the deadlock problem is asymmetrical: there is a philosopher who always takes the left `fork` first, there is also a philosopher who always takes the right `fork` first. The remaining philosophers take their `fork`s in arbitrary order. This solution requires two special servers: `chair_first_left` and `chair_first_right`, the other philosophers work as before. The code of the first of the special philosophers is below.

```
1. server:  chair_first_left(agents ph; servers fork[2]),
2. services {may_eat, may_think, may_take_second, may_release_second},
3. states   {think, eat, first_left, first_right},
4. actions {
5.        {ph.chair_first_left.may_think, chair_first_left.think} ->
            {ph.chair_first_left.may_think, chair_first_left.think},
6.        {ph.chair_first_left.may_think, chair_first_left.think} ->
            {ph.fork[1].take_first, chair_first_left.first_left},
7.        {ph.chair_first_left.may_take_second, chair_first_left. first_left} ->
            {ph.fork[2].take_second, chair_first_left.eat},
8.        {ph.chair_first_left.may_eat, chair_first_left.eat} ->
            {ph.chair_first_left.may_eat, chair_first_left.eat},
9.        {ph.chair_first_left.may_eat, chair_first_left.eat} ->
            {ph.fork[1].release_first, chair_first_left.first_left},
10.       {ph.chair_first_left.may_eat, chair_first_left.eat} ->
            {ph.fork[2].release_first, chair_first_left.first_right},
11.       {ph.chair_first_left.may_release_second, chair_first_left.first_left} ->
            {ph.fork[2].release_second, chair_first_left.think},
12.       {ph.chair_first_left.may_release_second, chair_
            first_left.first_right} ->
            {ph.fork[1].release_second, chair_first_left.think},
13. };
```

Separate *fork* types pose a problem: *fork*s should be specific to a particular pair of *chair*s due to types of formal parameter. This is solved by the introducing a special type *self*, which informs that the actual parameter can be of any type, provided that the sets of services in all actual parameters used for the formal parameter are identical. The definition of *fork* server is:

```
server:   fork(agents ph[2]; servers chair[2]:self),
```

The declaration of server variables and the initialization is as follows:

```
servers: fork[5],chair[3],chair_first_right,chair_first_left;
agents: ph[5];
```

```
1. init     -> {
2.          chair[1](ph[1],fork[1,2]).think,
3.          chair[2](ph[2],fork[2,3]).think,
4.          chair[3](ph[3],fork[3,4]).think,
5.          chair_first_left(ph[4],fork[4,5]).think,
6.          chair_first_right(ph[5],fork[5,1]).think,
7.          fork[1](ph[5,1],chair_first_right,chair[1]).free,
8.          fork[2](ph[1,2],chair[1],chair[2]).free,
9.          fork[3](ph[2,3],chair[2],chair[3]).free,
10.         fork[3](ph[3,4],chair[3],chair_first_left).free,
11.         fork[4](ph[4,5],chair_first_left,chair_first_right).free,
12.         ph[1].chair[1].may_think,
13.         ph[2].chair[2].may_think,
14.         ph[3].chair[3].may_think,
```

```
15.          ph[4].chair_first_left.may_think,
16.          ph[5].chair_first_right.may_think,
17. }.
```

Dedan reports deadlock freeness as expected. The second solution to the problem is the generalization of the first one: two *butlers* guard the *forks*, preventing to take five same-named *forks* as first (left ones or right ones). The butler is constructed just as a bounded buffer: when the maximum of requesting processes is reached, no granting action is prepared (the repeater *i* maximum value in line 7 is 4). The code of a *butler* type is:

```
1. server:  butler(agents ph[5]; servers chair[5]),
2. services {take, release},
3. states   {v0,v[4]}
4. actions  {
5.          <j=1..5>{ph[j].butler.take, butler.v0} ->
                     {ph[j].chair[j].may_take_first, butler.v[1]},
6. <i=1..3><j=1..5>{ph[j].butler.take, butler.v[i]} ->
                     {ph[j].chair[j].may_take_first, butler.v[i+1]},
7. <i=2..4><j=1..5>{ph[j].butler.release, butler.v[i]} ->
                     {ph[j].chair[j].aft_release, butler.v[i-1]},
8.         <j=1..5>{ph[j].butler.release, butler.v[1]} ->
                     {ph[j].chair[j].aft_release, butler.v0},
9. };
```

The *butler* contains the counter of the philosophers requesting the left *forks* or the right *forks*. The states of the *butler* are modeled as the state vector *v*, with 0th element separated as *v0* due to indexing from 1 in Dedan.

All the mentioned models of "five philosophers" (erroneous system, the two presented solutions and a third solution with taking "two or none" forks) are contained in the File EXAMPLES.zip, available at ("Dedan Examples," n.d.). All the examples are prepared in the source code of the server view, and converted by the Dedan program to the agent view. The source codes are included in files which names end with 'S'. The agent views of the examples are contained in files which names end with 'T'.

- phil3S.txt—3 philosophers. The Dijkstra's problem of dining philosophers, limited to 3 philosophers. The deadlock occurs.
- phil3S-asym.txt—3 philosophers, asymmetric solution. The solution based on the order of obtaining the forks: one philosopher takes the left fork first and one takes the right fork first.
- phil3S-2orNo.txt—3 philosophers, "all-or-nothing" solution. The solution based on taking two forks if both are available, on no fork if at least one fork is taken.

- phil3S-butlers.txt—3 philosophers, "butlers" solution. The two butlers are responsible for a prohibition of taking all three left forks as first and of taking all three right forks as first.

The models for 5 philosophers are verified entirely under Dedan (except for "butlers", which is verified for 4 philosophers). The models for a higher number of philosophers are verified using external model checker (Uppaal).

5.3 Intersection

The verification techniques can be used in modeling transport systems. The intersection is a typical example. The quarters of the intersection can be treated as resources, and a vehicle takes them one after another. The quarters *Q* and approaching fields *A* are coded with the cardinal directions. A vehicle leaving the intersection returns to the set of vehicles awaiting the appropriate approach field, for example a vehicle leaving QNW to the west enters the queue at AW. The topography is shown in Fig. 5.8.

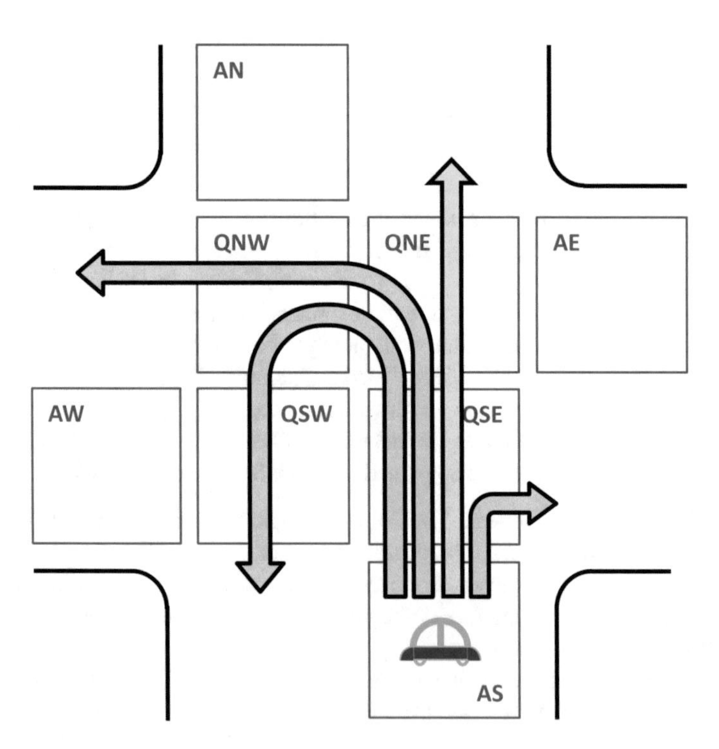

Fig. 5.8 Topography of intersection, Q-quarters, A-approaching fields, NSEW-directions (coded as cardinal directions)

Note that if a vehicle leaves the quarter, and waits for the next quarter freeing, other vehicles can do the same. This would lead to arising of a queue of vehicles between two neighboring quarters (which freed a quarter and request for the next one). The queue is simply the set of messages pending at the target quarter. Therefore, the vehicle must obtain the next quarter before freeing the current one, which prevents the queue from being created. The request message is issued to the next quarter, and if confirmation arrives, the vehicle moves. This requires three messages (*request, grant, move*), which form a kind of protocol of obtaining a quarter, for the example of QSE→QNE:

- QSE is occupied by the vehicle, and it wants to take QNE, The message `try` is sent from QSE to QNE. It is a *request*. This message may wait for acceptance for undefined period of time if QNE is occupied.
- If finally the message `try` is accepted in QNE (QNE is currently free), the message `ok` is sent back from QNE to QSE. It is a *grant*. QNE changes its state from `free` to `reserved`—it cannot be taken by other vehicles.
- Then, QSE is released and QNE is finally taken by the vehicle, QSE becomes `free` and QNE becomes `occupied`. It is a *move*.

We assumed a priority for the vehicles being on the intersection, i.e., QSE→QNE has priority over AE→QNE. The priority is realized by doubling the states `free`, `rese` and `occ` with their complementary states `freew`, `resew` and `occw` (*w* stands for *waiting*—line 5). The additional states denote a situation similar to the basic one, except that a vehicle with a priority wants to take the quarter and the vehicle without priority must wait. If a quarter is `free`—it is simply taken by a requesting vehicle. After the granting procedure begun, and the vehicle with priority requests the quarter —its state is changed to a complementary one (l.9), and after freeing the quarter it is granted immediately to a requesting vehicle (l.8). The protocol may seem to be complicated for the vehicle driver, but we can treat the system as the automatic vehicle guidance environment (see the next section), in which vehicles are conducted by field controllers cooperating with each other using the mentioned protocol.

The source code of the model is as follows:

```
1. #DEFINE N 4
2. #DEFINE K 4

3. server: quarter(agents car[N];
          servers prevq:quarter,nextq:quarter,in:road,out:road),
4. services{trye[K],oke[K],takee[K],tryi[K],takei[K],occupy[K]},
5. states   {free,freew,rese,resew,resi,occ,occw},

6. actions {
7. <i=1..N><j=1..K>   {car[i].quarter.tryi[j], quarter.free} ->
                           {car[i].in.oki[j], quarter.resi},
```

```
8. <i=1..N><j=1..K>  {car[i].quarter.tryi[j], quarter.freew} ->
                         {car[i].in.oki[j], quarter.resi},
9. <i=1..N><j=1..K>  {car[i].quarter.tryi[j], quarter.rese} ->
                         {car[i].quarter.tryi[j], quarter.resew},
10. <i=1..N><j=1..K>  {car[i].quarter.takei[j], quarter.resi} ->
                         {car[i].quarter.occupy[j], quarter.occ},

11. <i=1..N><j=1..K-1>{car[i].quarter.trye[j], quarter.free} ->
                         {car[i].prevq.oke[j+1], quarter.rese},
12. <i=1..N><j=1..K-1>{car[i].quarter.takee[j], quarter.rese} ->
                         {car[i].quarter.occupy[j], quarter.occ},
13. <i=1..N><j=1..K-1>{car[i].quarter.takee[j], quarter.resew} ->
                         {car[i].quarter.occupy[j], quarter.occw},

14. <i=1..N>          {car[i].quarter.occupy[1], quarter.occ} ->
                         {car[i].out.leave, quarter.free},
15. <i=1..N>          {car[i].quarter.occupy[1], quarter.occw} ->
                         {car[i].out.leave, quarter.freew},
16. <i=1..N><j=2..K>  {car[i].quarter.occupy[j], quarter.occ} ->
                         {car[i].nextq.trye[j-1], quarter.occ},
17. <i=1..N><j=2..K>  {car[i].quarter.occupy[j], quarter.occw} ->
                         {car[i].nextq.trye[j-1], quarter.occw},

18. <i=1..N><j=2..K>  {car[i].quarter.oke[j], quarter.occ} ->
                         {car[i].nextq.takee[j-1], quarter.free},
19. <i=1..N><j=2..K>  {car[i].quarter.oke[j], quarter.occw} ->
                         {car[i].nextq.takee[j-1], quarter.freew},
20. };

21. server:  road(agents car[N]; servers qin:quarter),
22. services {leave,oki[K]},
23. states   {idle,tries},
24. actions  {
25. //j=1 right, j=2 straight, j=3 left, j=4 turn back
26. <i=1..N><j=1..K> {car[i].road.leave, road.idle} ->
                       {car[i].qin.tryi[j], road.tries},
27. <i=1..N><j=1..K> {car[i].road.oki[j], road.tries} ->
                       {car[i].qin.takei[j], road.idle},
28. };

29. servers quarter[K],road[K];
30. agents  car[N];
```

```
31. init -> {
32.         quarter[1](car[1..N],quarter[4,2],road[1,2]).free,
33.         quarter[2](car[1..N],quarter[1,3],road[2,3]).free,
34.         quarter[3](car[1..N],quarter[2,4],road[3,4]).free,
35.         quarter[4](car[1..N],quarter[3,1],road[4,1]).free,
36. <j=1..4>road[j](car[1..N],quarter[j]).idle,

37. <i=1..N>car[i].road[1].leave,
38. }.
```

Verification shows a deadlock when all quarters are occupied and all vehicles request the next quarters. A deadlock does not occur if there are only three vehicles, or if all vehicles turn right.

This verification is rather simple and its results are obvious, but it shows the direction of research that can be carried out on transport systems, for example:

- guiding of vehicles in Personal Rapid Transport (PRT) station, equipped with stub-berths and a shunt, with a possibility of reversing or not (Daszczuk and Mieścicki 2014),
- railway interlocking system (Vu et al. 2015),
- railway control system (Nardone et al. 2016),
- etc.

5.4 Automatic Vehicle Guidance System

The analysis of vehicles (autonomous moving platforms—AMPs) behavior in the automatic guidance environment is analyzed in (Czejdo et al. 2016). The system topology is presented in Fig. 5.9. The solid lines are the track segments (the whole topography is letter E-shaped), and the boxes represent the distributed track segment controllers.

In the described system, we identify servers with static elements: the road segment controllers of the AMPs environment (warehouse lots and road markers). The controllers cooperate by means of simple protocols. The agents are identified with dynamic elements: AMPs traveling through the environment. For example, if the agent `AMP1` is in the place Road Marker `E1`, then it tries to take the Road Marker `M`. To do this safely, a protocol similar to that used in the intersection example is applied.

The views of the modeled system emphasize different aspects of the behavior, but it is worth stressing that they are merely two aspects of the uniform system. The views are simply different grouping of actions. The system in the server view allows the observation of the system from the perspective of road segments' controllers. The agent view gives the vehicles' perspective.

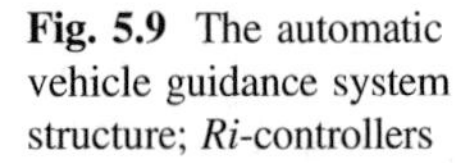
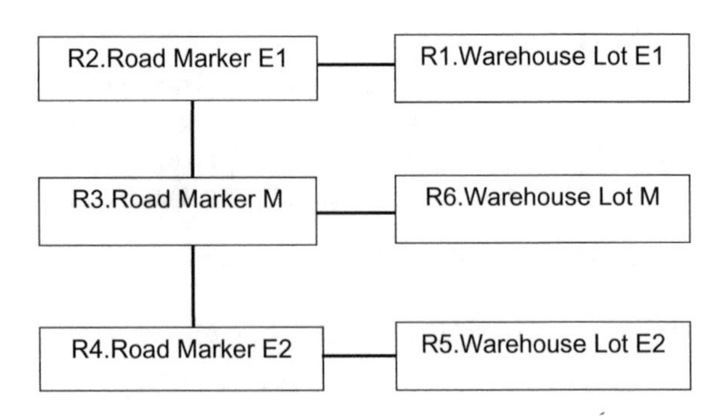

Fig. 5.9 The automatic vehicle guidance system structure; *Ri*-controllers

The operation of two vehicles is checked: one vehicle traveling from *Lot* `E1` to *Lot* `E2` and the other one from *Lot* `E2` to *Lot* `E1`. Proper behavior allows for bypassing using *Lot* `M1`. Indeed, the overtaking is performed correctly, but deadlocks can occur in road segments `R2` and `R4`, when one of the vehicles occupies the segment and the other one does not start driving yet.

Figure 5.10 shows the resource deadlock in the agent view.

5.5 Resource Deadlock Without Communication Deadlock

Typically a communication deadlock in the server view manifests as a resource deadlock in the agent view. However, in some cases of systems with too little resources, the server view can be free from communication deadlock, while some agents fall into the resource deadlock in the agent view (those for which there is a lack of resources). It is illustrated in Res_DeadlockS.txt in the examples available at ("Dedan Examples," n.d.). The system contains not enough resources. A resource deadlock occurs, because one of requesting agents waits forever for a resource, but there is no communication deadlock: all servers work, including the server containing the critical resource. This is due to the different formulations of communication deadlock and resource deadlock: communication deadlock affects all agents which messages are pending at the server, while resource deadlock may affect a single agent. The with lacking resources is stuck, while the other agents work, having allocated resources. Agents with insufficient resources fall into a deadlock individually, but checking for a total deadlock of all agents does not show any malfunction.

Such a situation may occur in the case of roundabout modeling: if all segments (quarters) of the roundabout are occupied, and the vehicles constantly omit one of the roads (from the direction E), vehicle approaching from this road become deadlocked. This is not a communication deadlock of the servers (segments of the roundabout), because the vehicles on these segments are moving. But in the agent view, we see the agents remaining on the bypassed road in a resource deadlock

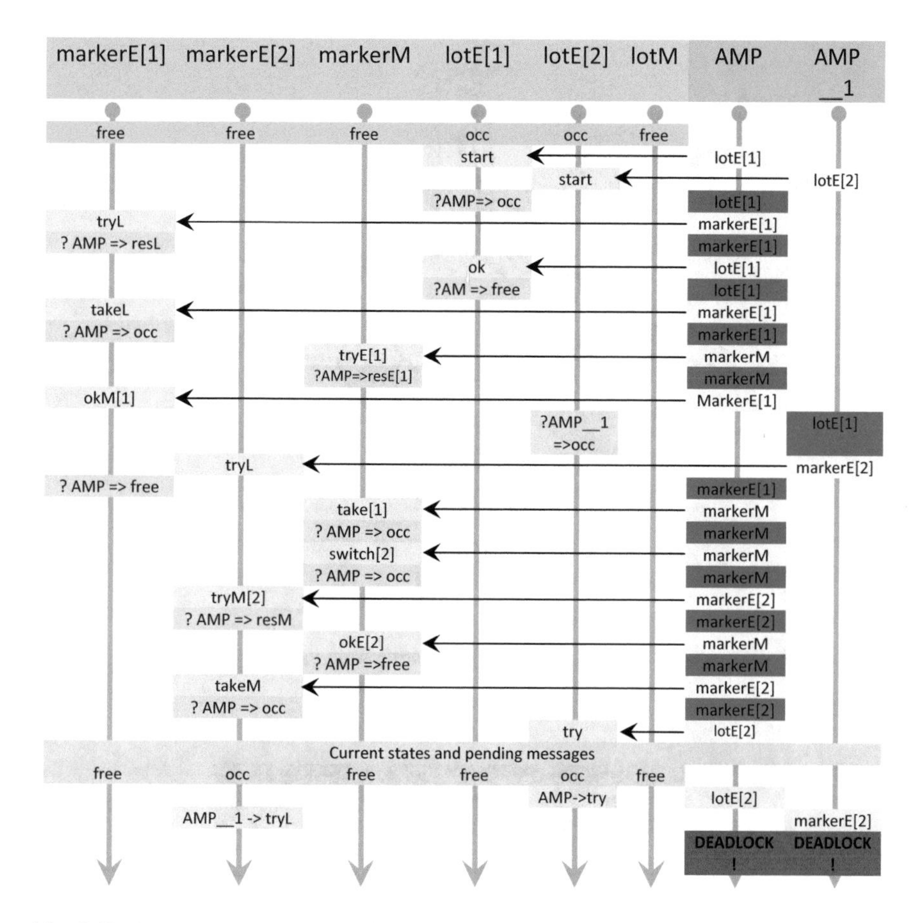

Fig. 5.10 Sequence diagram of *AMP* (AMP1) moving from Warehouse Lot *E1* to Warehouse Lot *E2* and *AMP__1* (AMP2) moving reverse way, leading to the deadlock

(Fig. 5.11). This is a deadlock in the sense of our definition: the agent's message is waiting for acceptance, but it will never be served. Such a situation, in which a single process can be blocked, is sometimes called a *starvation* (Tai 1994) or *stall* (Masticola and Ryder 1990).

5.6 Production Cell

The Production Cell is a real-world system developed in Karlsruhe. It was a benchmark for several research projects regarding the specification, design, synthesis, testing and verification of the controller for the Production Cell. The research is summarized in (Lewerentz and Lindner 1995). After the publication of the

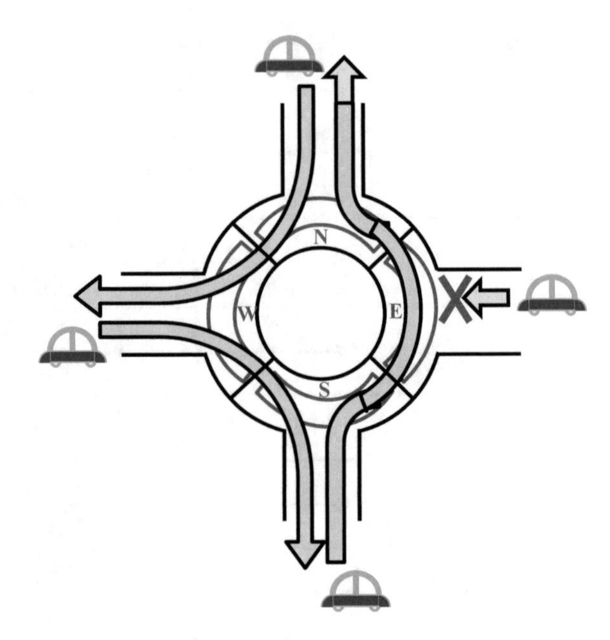

Fig. 5.11 Roundabout with blocking of vehicles entering from the right

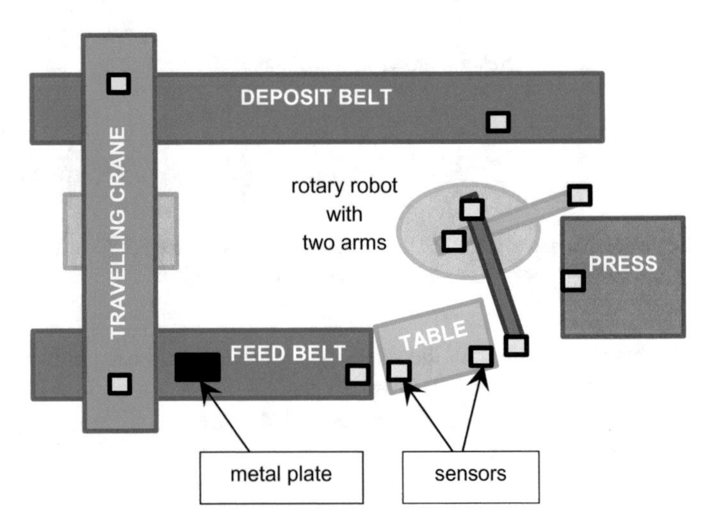

Fig. 5.12 Production cell

original book, many additional research papers about the benchmark were published. In several works Production Cell was used as an example among other benchmarks. Similar real-world systems were used in a number of papers.

The original benchmark (Fig. 5.12) is a set of devices for delivering metal plates (blanks) with a feed belt, placing them into the press by a two-armed rotary robot, withdrawing forged plates back and transferring them to the deposit belt. Equipping the robot with two arms allows simultaneous transport of two plates: the blank one to the press and the previously forged from the press. The difference in the levels of the feeder belt and the press is bridged by the elevating rotary table between them.

The Production Cell is the physical model, in which the press only pretends to forge the plates, so they remain blank and can be turned back in a loop. It is just the purpose of using the traveling crane which moves the plates between the belts.

5.6.1 Centralized Versus Distributed Modeling

Some approaches assume a centralized controller: Tatzelwurm (Klingenbeck and Käufl 1995) and TLT (Cuellar and Huber 1995). Most studies, however, concern distributed control of the Production Cell. The distribution is modeled as common variables between the subcontrollers (Heiner and Heisel 1999) or synchronous message passing (Greenyer et al. 2013; Garavel and Serwe 2017; Jacobs and Simpson 2015).

5.6.2 Synchrony Versus Asynchrony of Specification

Most models are based on synchronous paradigm, i.e., the devices agree their states on their joint activities (Greenyer et al. 2013; Garavel and Serwe 2017; Jacobs and Simpson 2015; Ma and Wonham 2005; Zorzo et al. 1999; Sokolsky et al. 1999; Ramakrishnan and McGregor 2000). Many of them are based on well-known formalisms: LOTOS (Rosa and Cunha 2004), CSP (Hoare 1978) or CCS (Milner 1984). Some papers deal with real time modeling (Cassez et al. 2005; Dierks 1996; Beyer et al. 2003; Burns 2003; Benghazi Akhlaki et al. 2007). Synchrony requires some kind of nonlocality, for example sensors common to the controllers of both neighboring devices.

Only a few papers deal with asynchronous modeling: SDL (Heinkel and Lindner 1995), provides asynchronous, simple protocols between devices. Focus (Fuchs and Philipps 1995; Barbey et al. 1998b) covers the asynchronous network of components (agents) that work concurrently and communicate over unbounded FIFO channels, translated into either a functional or an imperative programming language. It is admitted that to each arrow corresponds an asynchronous event, i.e., the event is sent even though the receiving agent is not ready to accept it. The Promela model in (Cattel 1995) uses asynchronous channels. The Model Checking Kit (Schröter et al. 2003) allows the preparation of models in various languages, including IF (Interchange Format): a language proposed for modeling asynchronous real-time communicating systems. Architecture Analysis and Design Language (Björnander et al. 2011) provides asynchronous specification, as the pairs of events `feedBelt.InFeedBeltReady->loader.InFeedBeltReady`.

Nonblocking Supervisory Control of State Tree Structures (Ma and Wonham 2005) provide Hierarchical Statecharts description. A synchronous system is modeled directly in BDD with a controlled order of BDD variables. Features are written as formulas over states, similar to invariants. A combination of AND-states (Cartesian products of component states) and OR-states (exclusive-or between component states) model asynchrony of structures, similar to messages pending at servers. We may say that AND states hierarchically model the structure and parallelism, while OR states model the behavior (the dynamics).

In some papers asynchronous communication is modeled by data structures: Multiparty Session Processes with Exceptions (Capecchi et al. 2016) use messages queues. In (Heiner and Heisel 1999) asynchrony is introduced as a place "in between" in a Petri net (1-element buffer).

5.6.3 Real-Time Modeling

Most papers on modeling the Production Cell deal with timeless control sequences. However, some of them deal with real-time constraints. In HTTDs [Hierarchical Timed Transition Diagrams (Cardell-Oliver 1995)], sequences of timed transitions are used. In (Ruf and Kropf 1999), Networks of Communicating Real-Time Processes are described: various levels of abstraction, I/O-interval structures and input-sensitive transitions. In (Benghazi Akhlaki et al. 2007), systematic transformation from UML–RT (Grosu et al. 1999) to CSP+T (time) (Žic 1994) is applied. Timed Game Automata (Ehlers et al. 2010) and Timed Games (Cassez et al. 2005) are used for on-the-fly verification. The timed specification language SL^{time} is described in (Dierks 1996), while graphical specification with time is possible in (Ben-Abdallah and Lee 1998). Timed Automata are used in (Burns 2003) and Cottbus Timed Automata in (Beyer and Rust 1998).

5.6.4 Message Passing/Resource Sharing

Synchronous modeling usually is based on variables, as in Lustre (Holenderski 1995), or synchronous channels (Greenyer et al. 2013; Garavel and Serwe 2017; Jacobs and Simpson 2015). Communication based on resource sharing in a distributed environment is presented in Graphical Communicating Shared Resources (Ben-Abdallah and Lee 1998).

5.6.5 Synthesis and Verification

Some papers concern the synthesis of Production Cell controller (or a set of distributed controllers of individual devices). Such are Esterel (Budde 1995), SDL (Heinkel and Lindner 1995), Deductive Synthesis using first order predicate logic (Burghardt 1995), Fusion method (Barbey et al. 1998b) using regular expressions. Modal Sequence

Diagrams and Feature Diagrams (Greenyer et al. 2013) are used for incremental synthesis exploiting the similarities between individual device specifications.

Testing and simulation as validation methods are addressed in several papers: Symbolic Timing Diagrams (Korf and Schlör 1995) are used for testing in waveform monitoring. Formal testing method is applied in (Barbey et al. 1998a) and (Larsen et al. 2009). Statistical testing of Reusable Concurrent Objects is described in (Waeselynck and Thévenod-Fosse 1999). A simulation is used as informal validation in SDL (Heinkel and Lindner 1995) and Graphical Communicating Shared Resources (Ben-Abdallah and Lee 1998).

In general, safety conditions as avoiding machine collisions and liveness properties are checked: if two blanks are inserted into the system, one of them will eventually arrive at the traveling crane.

5.6.6 Modeling Production Cell in IMDS

The IMS specification of Production Cell benchmark consists of the definition of servers (devices) and agents (blanks traveling through the cell). Every pair of servers negotiate using simple protocol, which is a sequence of three messages: the supplier send a message `try` that tests the receiver's readiness to accept the blank. When the receiver responds with a message `ok`, the supplier can deliver the blank. The supplier then issues a message `deliver`, which has two meanings: completion of the protocol and passing a blank from the supplier to the receiver. For example, the protocol between the rotary table (*T*) and the robot (*R*), viewed from the robot's perspective, has the following IMDS code:

```
<i=1..N>{A[i].R.tryIn, R.arm1_at_T} -> {A[i].T.ok, R.arm1_at_T},
<i=1..N>{A[i].R.deliverIn, R.arm1_at_T} -> {A[i].R. rotate_ccw_P, R. R.rot_ccw_a2_P},
```

The meaning of this specification is as follows: for each agent (`<i=1..N>`), if the message `tryIn` is received by the robot and its state denotes staying at receiving position (`arm1_at_T`), the message `ok` is issued and the robot remains in `arm1_at_T` state, waiting for the message `deliverIn`. When it arrives, the robot starts to rotate counterclockwise, issuing the message `rotate_ccw_P` to itself and entering the state `rot_ccw_a2_P`, which means that the robot rotates until its `arm2` points to the press (*P*). All the above operations are performed in the context of the agent `A[i]`, modeling the ith blank.

It is assumed that the belts do not stop, that's why they deliver blanks to receiving devices without using any protocol, they simply issue the message `deliver`.

Agents `A[1]..A[N]` model *N* blanks traveling through the cell. Sometimes, however, operations such as returning the table to `at_FB` position after delivering a blank to the robot are desirable. It is modeled by the additional agents internal to some devices, for example the agent `LT` in the table (*T*):

```
{LT.T.return, T.at_R_free} -> {LT.T.go_to_FB, T.mov_to_FB},
{LT.T.go_to_FB, T.mov_to_FB} -> {LT.FBC.ready, T.waits},
{LT.T.ok_ready, T.waits} -> {LT.T.return, T.at_FB},
```

The robot is equipped with two arms which extend for picking/leaving the blanks, and which retract before the robot rotation. For example, if the robot reaches the utmost position at the table and it receives the `tryIn` message, it extends to the appropriate length and then issues the message `ok`. After picking a blank, the arm retracts and the robot starts to rotate. This changes the previous specification (the "black box" symbol in the code ■ is described in the next sections):

```
<i=1..N>{A[i].R.tryIn, R.arm1_at_T} -> {A[i].R.wait_a1extIn, R.arm1_at_T_ext},
   {AARM[1].R.extend, R.arm1_at_T_ext} -> {AARM[1].ARM[1].extend, R.arm1_at_T_ext},
   {AARM[1].R.ok_ext, R.arm1_at_T_ext} -> {AARM[1].R.retract, R.arm1_at_T_long},
<i=1..N>{A[i].R.wait_a1extIn, R.arm1_at_T_long} -> {A[i].T.ok, R.arm1_at_T_long},
■<i=1..N>{A[i].R.deliverIn, R.arm1_at_T_long} ->
      {A[i].R.wait_a1retIn, R.arm1_at_T_ret},
   {AARM[1].R.retract, R.arm1_at_T_ret} -> {AARM[1].ARM[1].retract,R.arm1_at_T_ret},
   {AARM[1].R.ok_ret, R.arm1_at_T_ret} -> {AARM[1].R.extend, R.arm1_at_T_short},
<i=1..N>{A[i].R.wait_a1retIn, R.arm1_at_T_short} ->
      {A[i].R.rotate_ccw_P, R.rot_ccw_a2_P},
...
```

The complete model of the traveling crane (*C*) is as follows:

```
server: C(agents A[N],LC; servers DBH,FBH),
states {at_FB_occ,at_FB_free,mov_to_DB,at_DB,mov_to_FB,waits},
services {deliver,go_to_FB,go_to_DB,ok,return,ok_ready},actions{
<i=1..N>{A[i].C.deliver, C.at_DB} -> {A[i].C.go_to_FB, C.mov_to_FB},

<i=1..N>{A[i].C.go_to_FB, C.mov_to_FB} -> {A[i].FBH.try_C, C.at_FB_occ},
<i=1..N>{A[i].C.ok, C.at_FB_occ} -> {A[i].FBH.deliver_C, C.at_FB_free},
      {LC.C.return, C.at_FB_free} -> {LC.C.go_to_DB, C.mov_to_DB},
      {LC.C.go_to_DB, C.mov_to_DB} -> {LC.DBH.ready, C.waits},
      {LC.C.ok_ready, C.waits} -> {LC.C.return, T.at_DB},
}
```

5.6.7 Verification in Dedan

The critical situation occurs when the robot exchanges a fresh blank with a forged one in the press. Because it is forced to exchange the forged blank with the incoming, it is obvious that the last forged blank stays inside the press and cannot be taken out. If the circulation of blanks is applied in a closed loop, it requires at least two blanks for proper operation the system. Indeed, the verification shows a deadlock in the press, if only one blank is applied (Fig. 5.13, a large part of the sequence is omitted between the two black lines). The figure shows the server view, including the history of all servers. The last actions leading to a deadlock are enlarged in the lower right corner. These are messages exchanged between the robot (*R*) and the press (*P*).

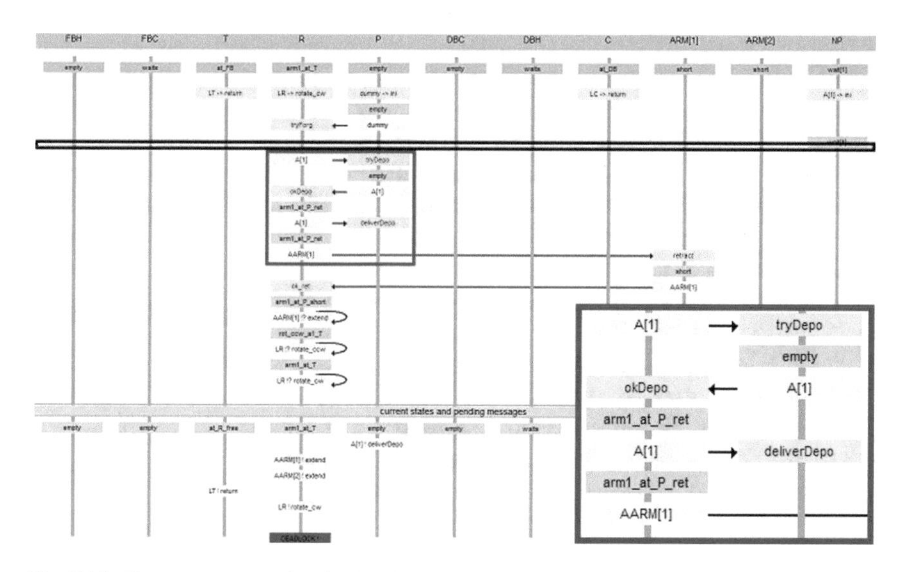

Fig. 5.13 The counterexample of a deadlock found in the server view

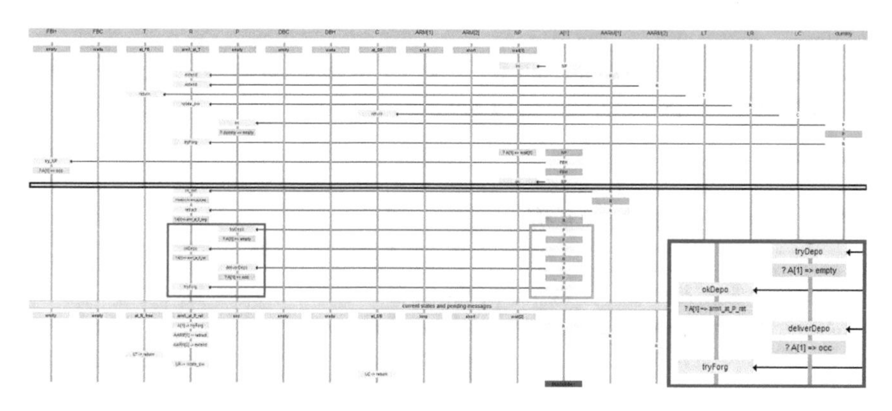

Fig. 5.14 The counterexample of a deadlock found in the agent view

The described situation is shown as a separate sequence of messages in a context of given agent `A[1]` in the agent view in Fig. 5.14. Last actions in red box on the left, enlarged in the lower right corner, leading to a deadlock in a context of the agent `A[1]` (green box on the right) are presented. Note that the server view shows the perspective of the cooperating controllers, and the deadlock applies to communication. The same system projected onto the agent view shows the perspective of blanks traveling through the cell, and shows the history that leads a blank to be stuck inside the press.

The correctness of the model was indicated for 2 and 3 blanks, and for 4 blanks the model exceeds the available memory (8 GB). Then, the external verifier Uppaal [with timeless modeling using urgent states, (Behrmann et al. 2006)] was used for 4 blanks and the verification showed correctness. The model with 5 blanks is too big for Uppaal, therefore a simplified specification was used. The robot arms were not modeled as separate servers and their activity was incorporated into the robot's operation:

```
<i=1..N>{A[i].R.tryIn, R.arm1_at_T} ->
          {A[i].R.wait_a1extIn, R.arm1_at_T_long},
<i=1..N>{A[i].R.wait_a1extIn, R.arm1_at_T_long} ->
          {A[i].T.ok, R.arm1_at_T_long},
<i=1..N>{A[i].R.deliverIn, R.arm1_at_T_long} ->
          {A[i].R.wait_a1retIn, R.arm1_at_T_short},
<i=1..N>{A[i].R.wait_a1retIn, R.arm1_at_T_short} ->
          {A[i].R.rotate_ccw_P, R.rot_ccw_a2_P},
```

This model was successfully verified for 5 blanks, but for 6 blanks the problem returned. Therefore, the next step was modeling the Production Cell as the open system, in which new blanks appear at the beginning of the feed belt and they disappear at the end on the deposit belt. The traveling crane is not used. This allowed to verify the system with 6 blanks (with a deadlock concerning the last blank, which is expected in the open loop).

5.6.8 Increasing Parallelism

When the robot reaches its target position, it extends one of its arms until the appropriate sensor stops its movement. Therefore, the arm cannot be extended in advance, before its direction points to the device. The same applies when the blank is picked/released: the arm retracts to its rear position. This is modeled as a set of actions, starting from the point marked with black box ■ in Sect. 5.6.6. When the robot reaches a state requiring retracting, a signal is issued to the arm, and after receiving of a confirmation, the robot begins to rotate, driven by the internal agent *LR*:

```
■<i=1..N>{A[i].R.deliverIn, R.arm1_at_T_long} ->
                {A[i].R.wait_a1retIn, R.arm1_at_T_ret},
     {AARM[1].R.retract, R.arm1_at_T_ret} ->
       {AARM[1].ARM[1].retract, R.arm1_at_T_ret},
     {AARM[1].R.ok_ret, R.arm1_at_T_ret} ->
       {AARM[1].R.extend, R.arm1_at_T_short},
```

```
<i=1..N>{A[i].R.wait_a1retIn, R.arm1_at_T_short} ->
                  {A[i].R.rotate_ccw_P, R.rot_ccw_a2_P},

            {LR.R.rotate_ccw, R.rot_cw_a2_P} -> {LR.R.
rotate_ccw, R.arm2_at_P},

<i=1..N>{A[i].R.rotate_ccw_P, R.arm1_at_P} ->
                {A[i].R.wait_a1extDepo, R.arm1_at_P_ext},
```

We noticed that the retracting differs from the extending of the arm: it does not depend on any sensor external to the robot. Instead, the internal robot sensor is used. Therefore, retracting the arm can be executed in parallel with the robot's rotation. This can be achieved by simply changing the position in code in which the retracted arm position is expected, and adding a new state `arm1_at_T_ret_goes` in which parallel actions occur:

```
■<i=1..N>{A[i].R.deliverIn, R.arm1_at_T_long} ->
                  {A[i].R.wait_a1retIn, R.arm1_at_T_ret},
          {AARM[1].R.retract, R.arm1_at_T_ret} ->
                    {AARM[1].ARM[1].retract, R.arm1_at_T_ret_goes},
<i=1..N>{A[i].R.wait_a1retIn, R.arm1_at_T_ret_goes} ->
                  {A[i].R.rotate_ccw_P, R.rot_ccw_a2_P},

          {LR.R.rotate_ccw, R.rot_cw_a2_P} -> {LR.R.rotate_ccw, R.arm2_at_P},

<i=1..N>{A[i].R.rotate_ccw_P, R.arm1_at_P} ->
                {A[i].R.wait_a1extDepo, R.arm1_at_P_wait},
    {AARM[1].R.ok_ret, R.arm1_at_P_wait} ->
    {AARM[1].R. extend, R.rot_ccw_a2_P_ext},
```

5.7 Time of Verification

Many examples were verified using Dedan, including several versions of bounded buffer (correct and deadlock-prone versions), dining philosophers, systems with resource allocation errors and others. These are examples of primarily didactic nature, with the time of generating reachability space of several seconds and verification time shorter than 1 s. Larger systems, such as the intersection in Sect. 5.3, with 8 servers, 4 agents and 752 actions, exceed the internal possibilities of Dedan due to the huge reachability space, impossible to build in the memory explicitly. However, the Sect. 5.3 example, exported to external model checker with symbolic space representation and CTL verification [Uppaal, (Behrmann et al. 2006)], gives reasonable export time of 65 s and verification time less than 1 s. The example reduced to 3 roads (6 servers, 3 agents, 306 actions) can be verified entirely under Dedan, with reachability space generation time about 8 h and verification time of several seconds. This confirms that with explicit reachability space representation, large examples cannot be checked. It will be better after space reduction and BDD symbolic representation techniques will be applied in Dedan (see Chap. 12). For now, large examples should be verified using external model checker, which can be invisible for a designer. An alternative method is a non-exhaustive verification using the "2-vagabonds" algorithm described in Chap. 11.

5.8 Use in Teaching—The Rybu Preprocessor

The Dedan program is used in ICS, WUT student laboratory. The students get assignments concerning synchronization examples and solve the problems using semaphores and monitors. After that, the students verify their solutions using Dedan (using the built-in verifier). A team of two students (Maciej Bielecki and Jan Michalski) elaborated a preprocessor called Rybu (Daszczuk et al. 2017), allowing the specification of a verified system in imperative-style language (Sebesta 1996, Chap. 2), defined by the authors. Rybu converts the system directly to the Dedan input file.

The main features of Rybu are:

- The syntax is C-like.
- The system is specified in the server view.
- The servers are divided into two kinds: reactive servers and processing servers.
- Reactive servers do not have initially assigned agents.
- The processing server has its own agent (exactly one) invisible in Rybu specification. The server contains a sequence of "instructions". The sequence terminates or forms a loop. Instructions form a sequence, just like in imperative languages, i.e., the successful completion of an instruction directs the "control" to the next one. To create a sequence in the Dedan code, the agent associated with the processing server sends a message to its home server, and the next instruction has this message on the input. Inside the instructions, messages are sent to the reactive servers and responses (messages as well) are obtained.
- A simple instruction consists of sending a message to a reactive server and acquiring a response, which can be *ok* or a value from a general enumeration type of possible reply values. The processing server is stateless, but local variables are planned in the final version of Rybu. Then, it will be possible to store a response in a local variable.
- The message sent by a processing server may contain a value of a type (see later).
- The choice instruction consists of a sending of message, obtaining a response and selecting one instruction from a set of nested instructions.
- The reactive server contains variables and state machine. Variables have the following types: enumeration, range, vector.
- Variables are converted to a Cartesian product of their values (as server states).

- Reactive server instructions simply change the server state based on the acquired messages, and send responses.
- The unconditional instruction is fired by any message; in this case the message matches all server states.
- A condition on the server state can be used together with an input message: only states that satisfy the condition are matched.
- The next message or the next state or both can be present on the output. No output state means that the state is preserved. If there is no output message, *ok* is sent.
- The output value of the server state may be ambiguous: many actions are generated that match this value. For example if one of several variables is assigned, all possible values of other variables are generated. The output value can be relative to an input state value (the expression is calculated). The choice between ambiguous output states in nondeterministic.
- For reactive servers, server types and server variables can be declared. Processing servers are variables, each declared separately. Processing server types are planned in the final version.

The Rybu program, together with several examples, is available at ("Rybu," n.d.). Examples include the "*vector*" problem (illustrating the variable vectors), the mid-semester test problem and the laboratory problem. All three examples concern buffers with complex put/get conditions. The solutions have 131, 59 and 129 lines of source code, respectively. The output code for Dedan consists of 5955, 84 and 179 lines of much denser code, respectively. The large Dedan code in the *vector* example is due to the explosion of values in the Cartesian product of vector elements.

Some students found deadlocks in their solutions, the other ones proved the correctness of their code. But both groups of students, using Rybu and using Dedan directly, appreciated lessons from the formal verification. However, there are still other opinions that confirm the misunderstanding of the nature of synchronization in concurrent systems and the significance of formal verification (translation by W. B. Daszczuk):

"My conclusions: Dedan deals only with small examples, and writing in Rybu using servers 'mutex' and 'semaphore' gives an additional overhead (but a more comfortable implementation). Even with more than a total of 4 robots [servers, comment by W.B. Daszczuk] Dedan (with code generated by Rybu) does not give advice [students do not use external verifiers, comment by W.B. Daszczuk]. A version without Rybu can do the job so for about 5 or 6 robots. Generally, I do not see a good future for Dedan. In practice, to check for deadlocks, it simply works by inserting a delay with random time between synchronization operations into a program written in C++."

Fortunately, such opinions are in the vast minority. Such an approach of testing instead of formal verification led to a famous priority inversion error in Pathfinder Mars explorer, caused in well-tested software (Pedersen et al. 2000; Priority Inversion and Deadlock. Whitepaper 2011).

References

Barbey, S., Buchs, D., & Péraire, C. (1998a). *A case study for testing object-oriented software: A production cell*. Url: http://citeseer.ist.psu.edu/viewdoc/summary?doi=10.1.1.46.460.

Barbey, S., Buchs, D., & Péraire, C. (1998b). *Modelling the production cell case study using the fusion method*. Lausanne, Switzerland. Url: https://infoscience.epfl.ch/record/54618/files/Barbey98-298..ps.gz.

Behrmann, G., David, A., & Larsen, K. G. (2006). *A tutorial on Uppaal 4.0*. Aalborg, Denmark. Url: http://www.it.uu.se/research/group/darts/papers/texts/new-tutorial.pdf.

Ben-Abdallah, H., & Lee, I. (1998). A graphical language for specifying and analyzing real-time systems. *Integrated Computer-Aided Engineering*, *5*(4), 279–302. Url: ftp://ftp.cis.upenn.edu/pub/rtg/Paper/Full_Postscript/icae97.pdf.

Benghazi Akhlaki, K., Capel Tuñón, M. I., & Mendoza Morales, L. E. (2007). A methodological approach to the formal specification of real-time systems by transformation of UML-RT design models. *Science of Computer Programming, 65*(1), 41–56. https://doi.org/10.1016/j.scico.2006.08.005.

Beyer, D., Lewerentz, C., & Noack, A. (2003). Rabbit: A tool for BDD-based verification of real-time systems. In *Computer Aided Verification, CAV 2003, Boulder, CO, 8–12 July 2003, LNCS 2725* (pp. 122–125). Heidelberg: Springer. https://doi.org/10.1007/978-3-540-45069-6_13.

Beyer, D., & Rust, H. (1998). Modeling a production cell as a distributed real-time system with cottbus timed automata. In H. König & P. Langendörfer (Eds.), *Formale Beschreibungstechniken für verteilte Systeme, 8. GI/ITG-Fachgespräch, Cottbus, Germany, 4–5 June 1998*. München, Germany: Shaker Verlag. Url: https://www.sosy-lab.org/~dbeyer/Publications/1998-FBT.Modeling_a_Production_Cell_as_a_Distributed_Real-Time_System_with.Cottbus_Timed_Automata.pdf.

Björnander, S., Seceleanu, C., … Pettersson, P. (2011). ABV—A verifier for the architecture analysis and design language (AADL). In *6th IEEE International Conference on Engineering of Complex Computer Systems, Las Vegas, NV, 27–29 April 2011* (pp. 355–360). IEEE. https://doi.org/10.1109/iceccs.2011.43.

Budde, R. (1995). Esterel. In C. Lewerentz & T. Lindner (Eds.), *Formal development of reactive systems. LNCS Vol. 891, chapt. 5* (pp. 75–100). Heidelberg: Springer. https://doi.org/10.1007/3-540-58867-1_49.

Burghardt, J. (1995). Deductive synthesis. In C. Lewerentz & T. Lindner (Eds.), *Formal development of reactive systems. LNCS Vol. 891, chapt. 17* (pp. 295–309). Heidelberg: Springer. https://doi.org/10.1007/3-540-58867-1_61.

Burns, A. (2003). How to verify a safe real-time system—The application of model checking and timed automata to the production cell case study. *Real-Time Systems, 24*(2), 135–151. https://doi.org/10.1023/A:1021758401878.

Capecchi, S., Giachino, E., & Yoshida, N. (2016). Global escape in multiparty sessions. *Mathematical Structures in Computer Science, 26*(02), 156–205. https://doi.org/10.1017/S0960129514000164.

Cardell-Oliver, R. (1995). HTTDs and HOL. In C. Lewerentz & T. Lindner (Eds.), *Formal development of reactive systems. LNCS Vol. 891, chapt. 15* (pp. 261–276). Heidelberg: Springer. https://doi.org/10.1007/3-540-58867-1_59.

Cassez, F., David, A., … Lime, D. (2005). Efficient on-the-fly algorithms for the analysis of timed games. In *16th International Conference on Concurrency Theory (CONCUR'05), San Francisco, CA, 23–26 Aug. 2005, LNCS 3653* (pp. 66–80). Heidelberg: Springer. https://doi.org/10.1007/11539452_9.

Cattel, T. (1995). Process control design using SPIN. In *Spin Workshop, Montreal, Canada, 16 Oct. 1995*. Url: http://spinroot.com/spin/Workshops/ws95/cattel.pdf.

Cuellar, J., & Huber, M. (1995). TLT. In C. Lewerentz & T. Lindne (Eds.), *Formal development of reactive systems. LNCS Vol. 891, chapt.* (pp. 151–169). Heidelberg: Springer. https://doi.org/10.1007/3-540-58867-1_53.

Czejdo, B., Bhattacharya, S., ... Daszczuk, W. B. (2016). Improving resilience of autonomous moving platforms by real-time analysis of their cooperation. *Autobusy-TEST, 17*(6), 1294–1301. Url: http://www.autobusy-test.com.pl/images/stories/Do_pobrania/2016/nr%206/logistyka/10_l_czejdo_bhattacharya_baszun_daszczuk.pdf.

Daszczuk, W. B. (2003). *Verification of temporal properties in concurrent systems*. Warsaw University of Technology. Url: https://repo.pw.edu.pl/docstore/download/WEiTI-0b7425b5-2375-417b-b0fa-b1f61aed0623/Daszczuk.pdf.

Daszczuk, W. B., Bielecki, M., & Michalski, J. (2017). Rybu: Imperative-style preprocessor for verification of distributed systems in the Dedan environment. In *KKIO'17—Software Engineering Conference, Rzeszów, Poland, 14–16 Sept. 2017*. Polish Information Processing Society. arXiv:1710.02722.

Daszczuk, W. B., & Mieścicki, J. (2014). Chapter 7.2 Principles of PRT network simulation. In W. Choromański (Ed.), *Ecomobility. Innovative and ecological transport means (in Polish)* (pp. 193–209). Wydawnictwa Komunikacji i Łączności. ISBN: 978-83-206-1953-9.

Dedan Examples. (n.d.). Url: http://staff.ii.pw.edu.pl/dedan/files/examples.zip.

Dierks, H. (1996). The production cell: A verified real-time system. In *4th International Symposium on Formal Techniques in Real-Time and Fault-Tolerant Systems FTRTFT 1996: Uppsala, Sweden, 9–13 Sept. 1996, LNCS 1135* (pp. 208–227). Heidelberg: Springer. https://doi.org/10.1007/3-540-61648-9_42.

Dijkstra, E. W. (1971). Hierarchical ordering of sequential processes. *Acta Informatica, 1*(2), 115–138. https://doi.org/10.1007/BF00289519.

Ehlers, R., Mattmüller, R., & Peter, H.-J. (2010). Combining symbolic representations for solving timed games. In K. Chatterjee & T. A. Henzinger (Eds.), *8th International Conference on Formal Modeling and Analysis of Timed Systems, FORMATS 2010, Klosterneuburg, Austria, 8–10 Sept. 2010, LNCS 6246* (pp. 107–121). Heidelberg: Springer. https://doi.org/10.1007/978-3-642-15297-9_10.

Fuchs, M., & Philipps, J. (1995). Focus. In C. Lewerentz & T. Lindner (Eds.), *Formal development of reactive systems. LNCS Vol. 891, chapt. 11* (pp. 185–197). Heidelberg: Springer. https://doi.org/10.1007/3-540-58867-1_55.

Garavel, H., & Serwe, W. (2017). The unheralded value of the multiway rendezvous: Illustration with the production cell benchmark. *Electronic Proceedings in Theoretical Computer Science, 244*, 230–270. https://doi.org/10.4204/EPTCS.244.10.

Greenyer, J., Brenner, C., ... Gressi, E. (2013). Incrementally synthesizing controllers from scenario-based product line specifications. In *Proceedings of the 2013 9th Joint Meeting on Foundations of Software Engineering—ESEC/FSE 2013, Sankt Petersburg, Russia, 18–26 Aug. 2013* (pp. 433–443). New York: ACM Press. https://doi.org/10.1145/2491411.2491445.

Grosu, R., Broy, M., ... Stefănescu, G. (1999). What is behind UML-RT? In H. Kilov, B. Rumpe, & I. Simmonds (Eds.), *Behavioral specifications of businesses and systems* (pp. 75–90). Boston, MA: Springer US. https://doi.org/10.1007/978-1-4615-5229-1_6.

Heiner, M., & Heisel, M. (1999). Modeling safety-critical systems with Z and petri nets. In M. Felici, K. Kanoun, & A. Pasquini (Eds.), *SAFECOMP '99 Proceedings of the 18th International Conference on Computer Safety, Reliability and Security, Toulouse, France, 27–29 Sept. 1999, LNCS Vol. 1698* (pp. 361–374). Heidelberg: Springer. https://doi.org/10.1007/3-540-48249-0_31.

Heinkel, S., & Lindner, T. (1995). SDL. In C. Lewerentz & T. Lindner (Eds.), *Formal development of reactive systems. LNCS Vol. 891, chapt. 10* (pp. 171–183). Heidelberg: Springer. https://doi.org/10.1007/3-540-58867-1_54.

Hoare, C. A. R. (1978). Communicating sequential processes. *Communications of the ACM, 21*(8), 666–677. https://doi.org/10.1145/359576.359585.

Holenderski, L. (1995). Lustre. In C. Lewerentz & T. Lindner (Eds.), *Formal development of reactive systems. LNCS Vol. 891, chapt. 6* (pp. 101–112). Heidelberg: Springer. https://doi.org/10.1007/3-540-58867-1_50.

Jacobs, J., & Simpson, A. (2015). A formal model of SysML blocks using CSP for assured systems engineering. In *Formal Techniques for Safety-Critical Systems, Third International Workshop, FTSCS 2014, Luxembourg, 6–7 Nov. 2014, Communications in Computer and Information Science 476* (pp. 127–141). Heidelberg: Springer. https://doi.org/10.1007/978-3-319-17581-2_9.

Klingenbeck, S., & Käufl, T. (1995). Tatzelwurm. In C. Lewerentz & T. Lindner (Eds.), *Formal development of reactive systems. LNCS Vol. 891, chapt. 14* (pp. 247–259). Heidelberg: Springer. https://doi.org/10.1007/3-540-58867-1_58.

Korf, F., & Schlör, R. (1995). Symbolic timing diagrams. In C. Lewerentz & T. Lindner (Eds.), *Formal development of reactive systems. LNCS Vol. 891, chapt. 18* (pp. 311–331). Heidelberg: Springer. https://doi.org/10.1007/3-540-58867-1_62.

Larsen, P. G., Fitzgerald, J. S., & Riddle, S. (2009). Practice-oriented courses in formal methods using VDM++. *Formal Aspects of Computing, 21*(3), 245–257. https://doi.org/10.1007/s00165-008-0068-5.

Lewerentz, C., & Lindner, T. (Eds.). (1995). *Formal development of reactive systems, LNCS 891*. Heidelberg: Springer. https://doi.org/10.1007/3-540-58867-1.

Ma, C., & Wonham, W. M. (2005). The production cell example. Chapter 5. In *Nonblocking Supervisory Control of State Tree Structures. Lecture Notes in Control and Information Science 317* (pp. 127–144). Heidelberg: Springer. https://doi.org/10.1007/11382119_5.

Masticola, S. P., & Ryder, B. G. (1990). Static infinite wait anomaly detection in polynomial time. In *1990 International Conference on Parallel Processing, Urbana-Champaign, IL, 13–17 Aug. 1990* (Vol. 2, pp. 78–87). University Park, PA: Pennsylvania State University Press. Url: https://rucore.libraries.rutgers.edu/rutgers-lib/57963/.

Milner, R. (1984). *A calculus of communicating systems*. Heidelberg: Springer.

Nardone, R., Gentile, U., … Mazzocca, N. (2016). *Modeling railway control systems in promela* (pp. 121–136). https://doi.org/10.1007/978-3-319-29510-7_7.

Pedersen, M. H., Christiansen, M. K., & Glæsner, T. (2000). *Solving the priority inversion problem in legOS*. Aalborg, Denmark. Url: http://www.it.uu.se/edu/course/homepage/realtid/p2ht08/lego/prioinvers.pdf.

Priority Inversion and Deadlock. Whitepaper. (2011). Url: https://code-time.com/pdf/Priority%20Inversion%20and%20Deadlock.pdf.

Ramakrishnan, S., & McGregor, J. (2000). Modelling and testing OO distributed systems with temporal logic formalisms. In *8th International IASTED Conference Applied Informatics' 2000, Innsbruck, Austria, 14–17 Feb. 2000*. Url: https://research.monash.edu/en/publications/modelling-and-testing-oo-distributed-systems-with-temporal-logic-.

Rosa, N. S., & Cunha, P. R. F. (2004). A software architecture-based approach for formalising middleware behaviour. *Electronic Notes in Theoretical Computer Science, 108,* 39–51. https://doi.org/10.1016/j.entcs.2004.01.011.

Ruf, J., & Kropf, T. (1999). Modeling and checking networks of communicating real-time processes. In L. Pierre & T. Kropf (Eds.), *CHARME'99, Advanced Research Working Conference on Correct Hardware Design and Verification Methods, BadHerrenalb, Germany, 27–29 Sept. 1999, LNCS 1703* (pp. 267–279). Heidelberg: Springer. https://doi.org/10.1007/3-540-48153-2_20.

Rybu. (n.d.). https://zyla.neutrino.re/rybu/.

Schröter, C., Schwoon, S., & Esparza, J. (2003). The model-checking kit. In *24th International Conference ICATPN 2003: Eindhoven, The Netherlands, June 23–27, 2003, LNCS 2697* (pp. 463–472). Heidelberg: Springer. https://doi.org/10.1007/3-540-44919-1_29.

Sebesta, R. W. (1996). *Concepts of programming languages*. Reading, MA: Addison-Wesley Publishing Compan.

Sokolsky, O., Lee, I., & Ben-Abdallah, H. (1999). *Specification and analysis of real-time systems with PARAGON (equivalence checking)*. Philadelphia, PA. Url: https://www.cis.upenn.edu/~sokolsky/ase99.pdf.

Tai, K. (1994). Definitions and detection of deadlock, livelock, and starvation in concurrent programs. In D. P. Agrawal (Ed.), *1994 International Conference on Parallel Processing (ICPP'94), Raleigh, NC, 15–19 Aug. 1994* (pp. 69–72). Boca Raton: CRC Press. https://doi.org/10.1109/icpp.1994.84.

Vu, L. H., Haxthausen, A. E., & Peleska, J. (2015). *Formal modeling and verification of interlocking systems featuring sequential release* (pp. 223–238). https://doi.org/10.1007/978-3-319-17581-2_15.

Waeselynck, H., & Thévenod-Fosse, P. (1999). A case study in statistical testing of reusable concurrent objects. In J. Hlavička, E. Maehle, & A. Pataricza (Eds.), *Third European Dependable Computing Conference, Prague, Czech Republic, 15–17 Sept. 1999, LNCS 1667* (pp. 401–418). Heidelberg: Springer. https://doi.org/10.1007/3-540-48254-7_27.

Žic, J. J. (1994). Time-constrained buffer specifications in CSP+T and timed CSP. *ACM Transactions on Programming Languages and Systems, 16*(6), 1661–1674. https://doi.org/10.1145/197320.197322.

Zorzo, A. F., Romanovsky, A., ... Welch, I. S. (1999). Using coordinated atomic actions to design safety-critical systems: a production cell case study. *Software: Practice and Experience, 29*(8), 677–697. https://doi.org/10.1002/(sici)1097-024x(19990710)29:8%3c677::aid-spe251%3e3.0.co;2-z.

Chapter 6
Using the Dedan Program

6.1 Program Structure

The structure of the Dedan program is presented in Fig. 6.1. The central module Model contains the representation of a verified system. This module converts between the server view and the agent view of the system. The Input module allows reading the model from the file in IMDS notation or XML file. The Output module saves the model in various formats:

- IMDS (.TXT),
- Promela (.PML—for verification under Spin),
- SMV (.SMV—for verification under NuSMV),
- XML Timed Automata (.XML—for verification under Uppaal),
- ANDL (.ANDL—for analysis with Charlie),
- Petri net (.TXT—for analysis with other Petri net tools).

In the LTS module, the global configuration space is constructed. The purpose of this module is:

- temporal verification of deadlocks and termination (model checking, using TempoRG model checker),
- counterexample (trail) graphical display in the form of state and message sequence or configuration sequence,
- simulation over states and messages,
- simulation over LTS.

The Edition module allows building or changing the structure of the model. Experience shows that editing the IMDS source text file is much more convenient. The DA^3 (distributed automata are defined in Chap. 8) module supports graphical edition of the server/agent automata, and simulation over automata. This does not require generating the LTS.

© Springer Nature Switzerland AG 2020

W. B. Daszczuk, *Integrated Model of Distributed Systems*, Studies in Computational Intelligence 817, https://doi.org/10.1007/978-3-030-12835-7_6

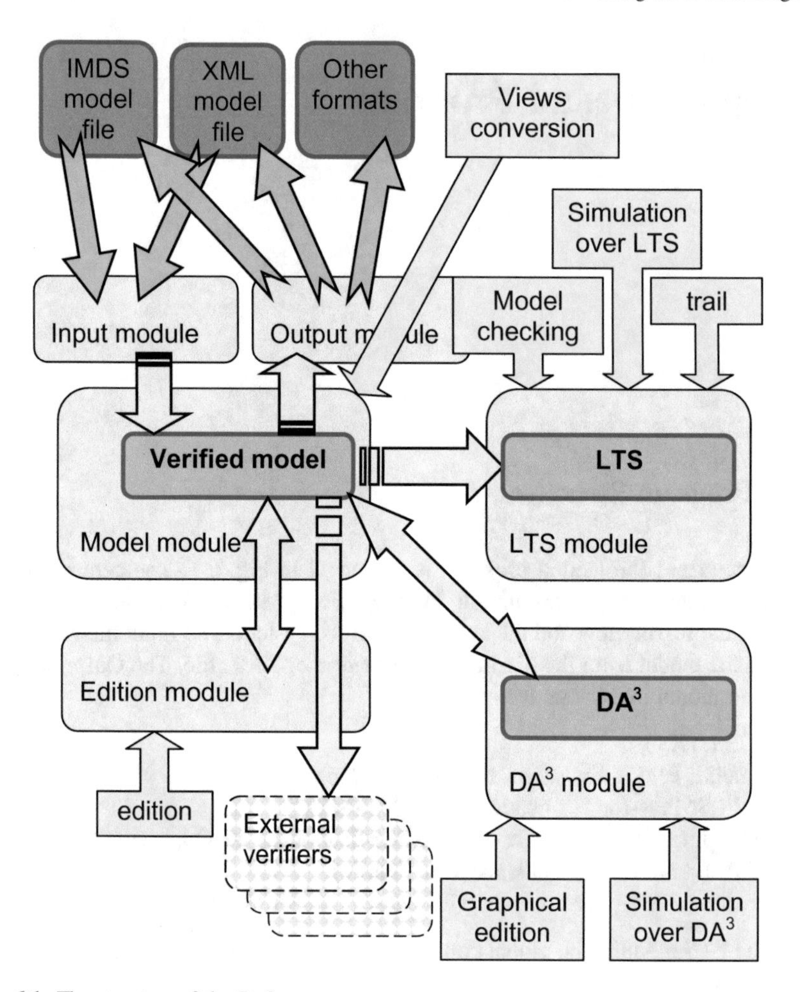

Fig. 6.1 The structure of the Dedan program

6.2 Main Window

The Dedan program, after its invoking, shows the set of tables for system specification in the server view (Edition module—Fig. 6.2). The server types are on the left while the agent types are on the right. The four middle tables contain the attributes of the selected server type: the states, services and formal parameters (used servers and used agents). Only these states, services and formal parameters can be used in the actions of the selected server type.

The list of actions in the server type is located at the bottom of the window. The action may be prefixed by repeaters (up to three). The repeater is an integer variable with a defined range. The repeaters are used in indices of the action elements. If the

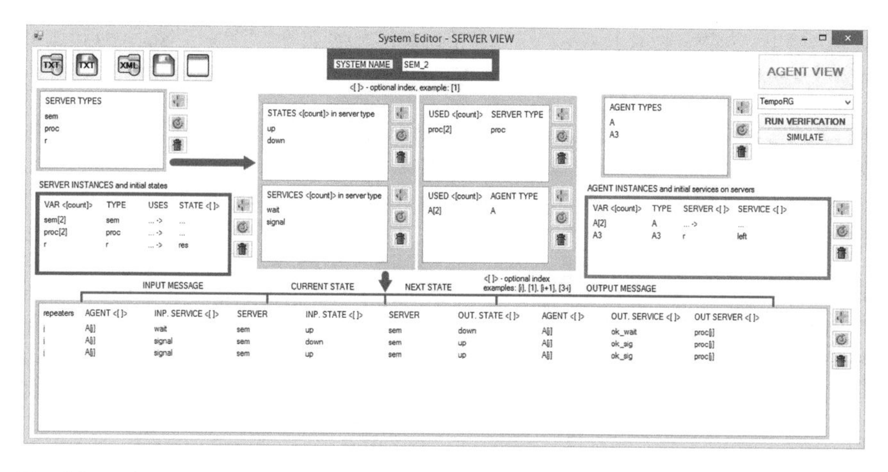

Fig. 6.2 Main window of the Dedan program

repeaters are used, then actions are generated for each vector in the Cartesian product of repeaters' values.

The action consists of the input message (agent, server and service), the input state and the output state. The output message is optional: it does not occur in the agent-terminating action. Repeaters can be used in the indices of servers, agents, states and services (if they are vectors). The indices may have the form of expressions: "repeater ± constant" or "constant ± repeater".

On the left, below the server types, server variables are listed. For each server variable, its type is given, and a "pointer" to the list of actual parameters (invoked by double clicking the pointer) and the initial state of the server variable. The window is presented in Fig. 6.3 on the left.

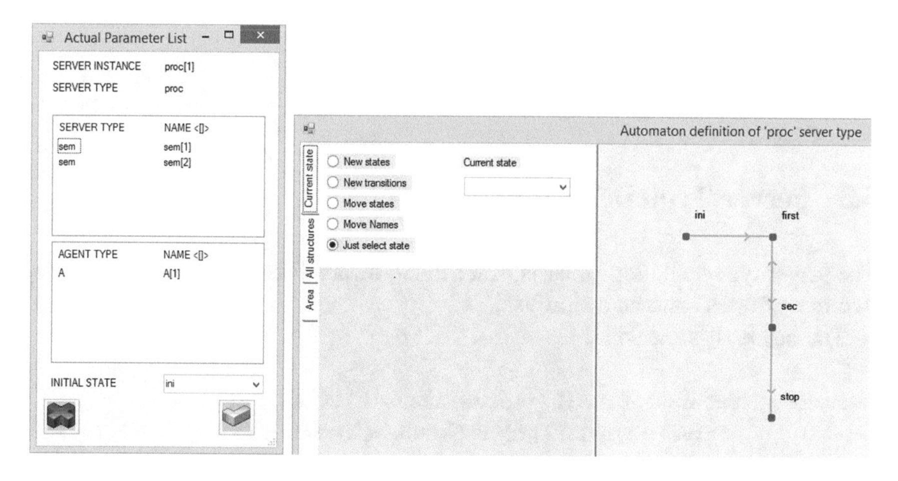

Fig. 6.3 Definition of actual parameters and initial state of a server (left) and graphical server design (right)

If the server variable is a vector, the pointer leads to the list of instances instead of the list of actual parameters. For each instance, its actual parameters and an initial state is defined.

On the right, below the agent types, agent variables are listed. For every agent variable, its initial message is defined (server and service).

Each server element: a state, service or a formal parameter that is a server or an agent—it can be a vector.

All lists allow for edition of their elements, as well as adding/removing elements. However, from the author's experience, editing the source IMDS text file is more convenient. The alternative method of definition is graphical specification of the server type automaton in DA^3 module (DA^3 distributed automata are defined in Chap. 8). The server automaton design window is presented in Fig. 6.3 on the right.

6.3 Files

The Input module and Output module provide access to model files. The file navigation buttons at the upper left corner are Open and Save text files in IMDS notation. The file in the server view is divided into the model name, server type definitions, optional agent type definitions, variable declarations and initialization part. The formal syntax is given in Appendix C.

6.4 Model Name

Optional model name is defined in the syntax:

```
system              BUF;
```

6.5 Server Type Definition

The server type definition consists of a header, a state list, a service list (the latter two in any order) and an action list.

The header has the form:

```
server: buf(agents Aprod[2]:Aprod,Acons[2]:Acons;
            servers Sprod[2]:Sprod,Scons[2]:Scons),
```

where *buf* is the server type identifier, and the list of used agents and the list of used servers follow in parentheses. These are formal parameters, the actual parameters are given in the initialization part. If the formal parameter types are identical to parameter names, the types can be omitted:

```
server: buf(agents Aprod[2],Acons[2]; servers Sprod[2],Scons[2]),
```

Sometimes it is useful to have parameters of different types at the same position in various instances of the same server type. In such a case, the type tag *self* can be used. The meaning of a parameter type *self* is "do not check the compatibility of types":

```
server: buf(agents Aprod[2],Acons[2];
        servers Sprod[2]:self,Scons[2]:self),
```

States and services are simply listed in the relevant lists:

```
states   {elem0,elem[3]} ,
services {  put, get} ,
```

Note that lists elements can be vectors (*elem[3]* in this example). Vectors are indexed from 1 to a maximum value (here, 3). This is the reason why the 0th element is excluded as a separate state in examples.

The actions are listed as pairs *{message, state} -> {next message, next state}* and may be prefixed by up to three repeaters, for example:

```
actions{
<i=1..3>             {Aprod[i].buf.put, buf.elem0} ->
                         {Aprod[i].Sprod[i].ok_put, buf.elem[1]},
...
```

The complete definition of the example server type is as follows:

```
server:          Sprod(agents Aprod;servers buf),
services          {doSth,ok_put},
states           {neutral,prod},
actions          {
        {Aprod.Sprod.doSth, Sprod.neutral} -> {Aprod.buf.put, Sprod.prod},
        {Aprod.Sprod.ok_put, Sprod.prod} -> {Aprod.Sprod.doSth, Sprod.neutral},
};
```

6.6 Agent Type Definition

The agent definition simply gives its name:

agent Aprod;

If an agent instance name (simple or vector) is identical to the agent type name, then the type need not be defined.

6.7 Instances Declaration

The instances of server types and agent types are declared as simple or vectors:

```
servers    buf,Sprod[3],Scons[3];
agents     Aprod[3],Acons[3];
```

6.8 Initialization Part

The initialization part begins with "***init*** *-> {*"and ends with "*}*". Server instances and agent instances are initialized in any order. The server instance must be supplied with a list of actual parameters (agent instances and server instances) and the initial state:

```
Sprod(Aprod,buf).neutral,
```

The initialization of the server instance or the agent instance as an element of the vector:

```
Sprod[1](Aprod[1],buf).neutral,
```

A repeater can be applied to initialize a vector:

```
<j=1..3 > Sprod[j](Aprod[j],buf).neutral,
```

If the actual parameters are vector elements, then they can be supplied individually (as above), or a list of indices or a range of indices can be applied:

```
buf(Aprod[1],Aprod[2],Acons[1,2],Sprod[1..2],Scons[1..2]).elem0,
```

The agent instance must be supplied witch an initial message, i.e., a service of a server:

```
Aprod.Sprod.doSth,
```

6.9 Macrogenerator

The macrogenerator allows defining integer constants, which can be used in index expressions:

```
#DEFINE N 3
#DEFINE K 2
server:              buf(agents Aprod[N],Acons[N]; servers Sprod[N],Scons[N]),
states               {elem0,elem[K]},
...
<i=1..N><j=1..K-1>          {Aprod[i].buf.put, buf.elem[j]} ->
                               {Aprod[i].Sprod[i].ok_put, buf.elem[j+1]},

init -> {
            buf(Aprod[1..N],Acons[1..N],Sprod[1..N],Scons[1..N]).elem0,
<j=1..N>   Aprod[j].Sprod[j].doSth,
```

6.10 Verification

The verification is performed in a window invoked by "RUN VERIFICATION" button. The verification window of LTS module is invoked. Five versions of the verification are developed (a list of verification environments is below the mentioned button):

- TempoRG—the verifier developed by the author of this monograph, adapted for the purpose of using it under the Dedan program. Adaptation consists in evaluating only a limited set of formulas, shown in Table 4.1, which speeds up the evaluation. Reverse reachability is used to identify the structure of the reachability space, see Sect. 9.5. TempoRG performs CTL verification (Clarke et al. 1999) on explicit reachability space, therefore large models cannot be verified. As the complete model is constructed, it can be observed as a whole (full Labeled Transition System) or simulated. The window is presented in Fig. 6.4.

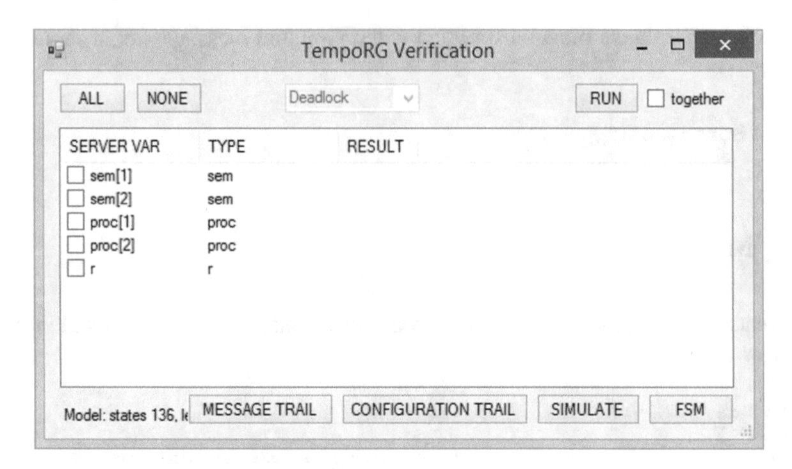

Fig. 6.4 Verification window

Deadlock detection can be performed for individual servers or for all selected servers together. In some systems, a subsets of the servers may fall into a deadlock (for example, individual server or server pair), but all servers in the set never fall into a common deadlock.

- Spin—the LTL verifier, requires installation of Spin (Holzmann 1995) and the GNU C compiler (32-bit Cygwin is recommended). Paths to Spin and C Compiler must be provided. Model preparation and verification can be performed as automatic sequence of program calls or step-by-step.
- SMV (connection in development)—the CTL and LTL verifier, based on the symbolic reachability space representation.
- Uppaal—TCTL (CTL with time constraints (Alur and Dill 1994)), based on symbolic reachability space representation. The connection Dedan → Uppaal works, the connection in opposite direction for interpretation of counterexamples is under development.
- 2-vagabonds—non-exhaustive verifier (see Chap. 11).

The counterexample consists of a sequence of messages sent between servers, and the states of these servers. At the end of the sequence, states and pending messages of the erroneous configuration are shown, for instance in Figs. 5.2 and 9.2.

The counterexample can also be observed as a system configuration sequence (Fig. 6.5). The configuration is shown textually on the left, while messages pending at individual servers are shown graphically on the right. The matching pair that enables the action is connected by a segment.

Counterexamples can be written to the file in formats: JPG, PNG, BMP, XML and PDF.

The verified system can be simulated as presented in Fig. 6.6. All matching pairs are connected by the arrows leading to a common box describing an action to

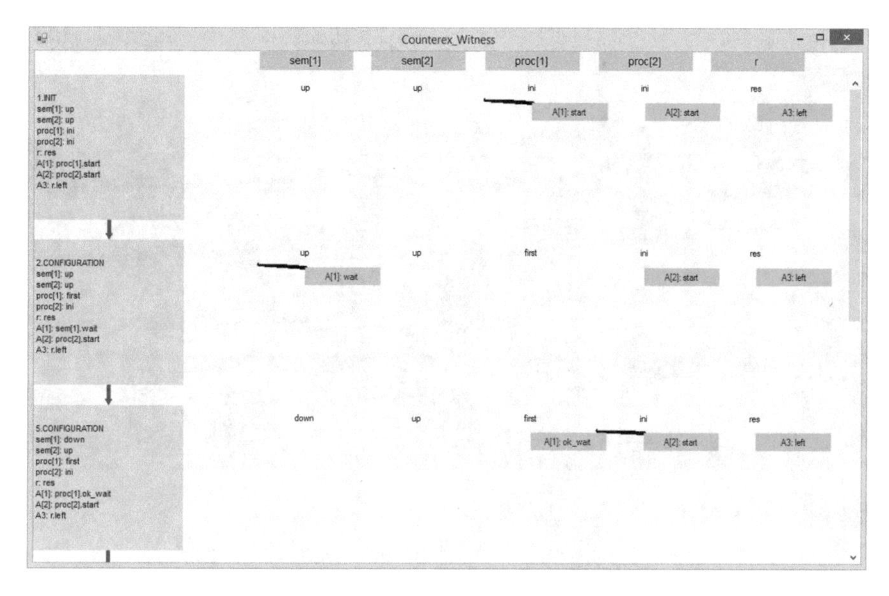

Fig. 6.5 Counterexample viewed as configuration sequence

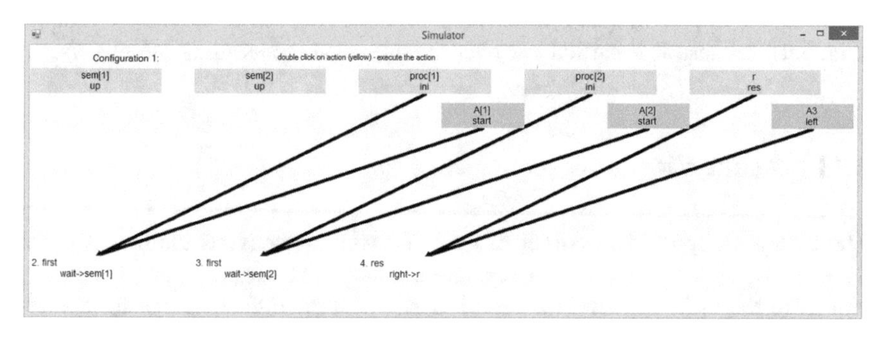

Fig. 6.6 Simulation of a model

execute: the next server state and the next agent message (if any). Clicking this box causes the execution of the action and the output configuration becomes current.

A entire LTS can be observed as a graph (Fig. 6.7). Simulation is possible over LTS. On the left, the current configuration is shown textually and a list of next configurations is listed, each with the matching input pair of the action, leading to the next configuration. Clicking one of the next configurations makes it current. On the right, each configuration is shown in a box, and possible next configurations are appointed by arrows. The current configuration is distinguished with a green box, and the red arrows lead to the next configurations. Matching pairs are displayed on the top of each possible next configuration on a pink background. Clicking any

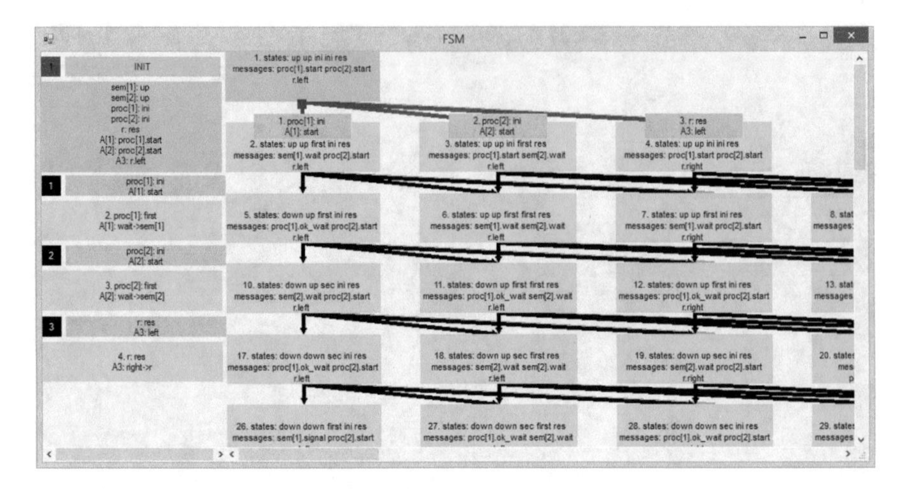

Fig. 6.7 LTS window

visible configuration results in a detailed display of its data and enables taking this configuration as current.

Simulation is possible without the calculating the global LTS, using the SIMULATE button in main window. This simulation is described in Chap. 8 (Fig. 8.4), because it is carried out over the automata representing the servers.

6.11 Agent View

The button on upper right corner of the main window converts the model to the agent view, and displays it in a separate window. The shape of this window is similar to the main window (the server view), with the difference that formal and actual parameters are attributed to the agent types and agent variables rather than to servers. Of course, states and services are still attributed to server types. It is shown in Fig. 6.8.

During the conversion process, agent vectors may be split into individual variables (when vector elements differ, for example in the lists of actual parameters). Verification is performed for agent processes. The difference lies in the deadlock definition: the agent process is deadlocked if its message is pending (at any server) and will never be accepted. In addition, the termination of the agent processes can be checked in the agent view. In the counterexample (or the witness) in the agent view, time lines of individual servers and individual agents are displayed.

Counterexamples in the agent view are presented in Figs. 5.4, 5.5 and 5.10.

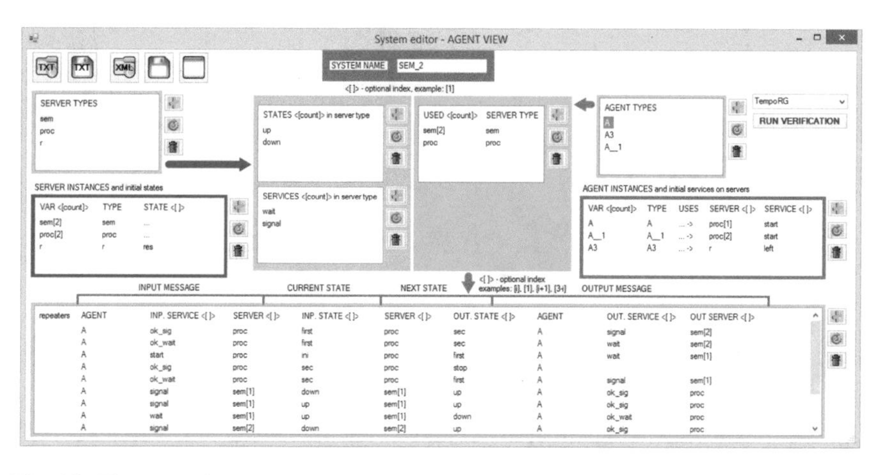

Fig. 6.8 The agent view

References

Alur, R., & Dill, D. L. (1994). A theory of timed automata. *Theoretical Computer Science, 126*(2), 183–235. https://doi.org/10.1016/0304-3975(94)90010-8.

Clarke, E. M., Grumberg, O., & Peled, D. (1999). *Model checking*. Cambridge, MA: MIT Press. ISBN: 978-0-262-03270-4.

Holzmann, G. J. (1995). Tutorial: Proving properties of concurrent systems with SPIN. In *6th International Conference on Concurrency Theory, CONCUR'95, Philadelphia, PA* (pp. 453–455), 21–24 August, 1995. Berlin, Heidelberg: Springer. https://doi.org/10.1007/3-540-60218-6_34.

Chapter 7
Deadlock Detection in Petri Net Equivalent to IMDS

7.1 Petri Net Equivalent to IMDS

Although Dedan's main goal is finding deadlocks, the user may be interested in the other properties of the verified system, for example:

- structural features of the system: structural conflicts, dead code (in Dedan it is a set of never-accessible actions), pure cyclic system or not, etc.,
- temporal properties other than deadlock: if the system is safe from some erroneous situation, if given situations are inevitable, etc.,
- graphical definition of concurrent system components (servers or agents),
- graphical simulation over concurrent components rather than in terms of a global graph.

In order to support these possibilities, several additional amenities are added to Dedan. In addition to exports to external model checkers, the interface with Charlie Petri net analyzer (Charlie, n.d.; Heiner et al. 2015) is added for structural analysis. The export is in ANDL format (*Abstract Net Description Language* (Schwarick 2013)).

The IMDS system can be converted to an equivalent Petri net (Daszczuk and Zuberek 2018). Each server state p is converted to a place p and each message m is converted to a place m. Each action (m, p) Λ (m', p') is equivalent to the Petri net (PN) transition λ, which has the places p and m on input and places p' and m' on output. In the case of the agent-terminating action (m, p) Λ (p'), the PN transition λ has only one output place p'. It is illustrated in Fig. 7.1. The colors in the figure are used for readability, it is not a colored Petri net in the sense of Jensen and Kristensen (2009).

The initial marking of the Petri net has tokens in all places of the initial server states and all places of the initial agent messages. By construction (of the described conversion of the IMDS system to a Petri net) the reachable markings graph of PN has identical structure as the LTS of IMDS: states $\leftrightarrow$ "red" places,

© Springer Nature Switzerland AG 2020

W. B. Daszczuk, *Integrated Model of Distributed Systems*, Studies in Computational Intelligence 817, https://doi.org/10.1007/978-3-030-12835-7_7

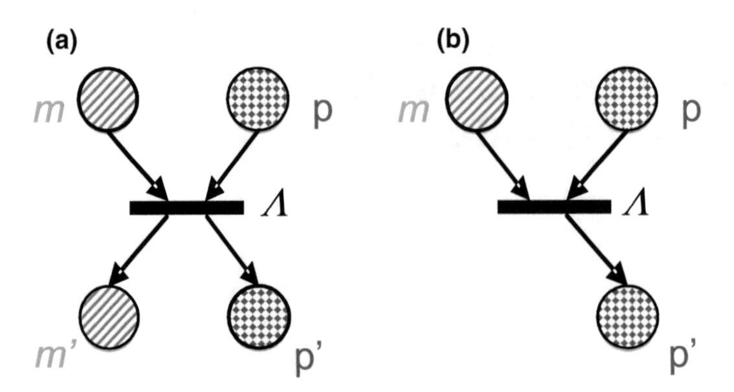

Fig. 7.1 Petri net interpretation of **a** regular action; **b** agent-terminating action

messages ↔ "green" places, actions ↔ PN transitions, configuration ↔ PN marking, initial configuration ↔ initial PN marking.

7.1.1 Petri Net Definition

Below we present the definition of a safe Petri net (PN with place capacity 1). It differs from literature (Zuberek 2009) in joining the definitions of transitions and arcs to a compound definition of transitions, which is more suitable for showing the correspondence between IMDS and PN.

- Petri net $= (H_{PN}, H_{PN0}, Tr)$, (7.1)

where:

  - H_{PN} is the set of *places*,
  - $H_{PN0} \subseteq H_{PN}$ is the set of *initial places*,
  - $Tr \subseteq 2^{H_{PN}} \times 2^{H_{PN}}$—the set of *transitions*, the transition $tr = (H_{inp}, H_{out}) \in Tr$ has a set of *input places* $H_{inp}(tr)$ and a set of *output places* $H_{out}(tr)$.

- $mark \subseteq H_{PN}$—*marking*—the subset of places which are marked, we say that each place in the marking is equipped with a *token.*
- $mark_0 = H_{PN0}$—the *initial marking*, i.e., the set of places equipped initially with tokens.
- Firing a transition: if $\exists_{tr \in Tr}$ $tr = (H_{inp}, H_{out})$, $mark \cap H_{inp}(tr) = H_{inp}(tr) \wedge mark \cap H_{out}(tr) \backslash H_{inp}(tr) = \varnothing \Rightarrow mark' = mark \backslash H_{inp}(tr) \cup H_{out}(tr)$, $H_{inp}(tr)$ is the *input set* of *tr* and $H_{out}(tr)$ is the *output set* of *tr*; *mark* is an *input marking* $mark_{inp}(tr) \supseteq H_{inp}(tr)$ and *mark'* is an *output marking*

$mark_{out}(tr) \supseteq H_{out}(tr)$; we say that the transition removes input tokens from its input set and inserts output tokens into its output set.
The transition tr is *enabled* in $mark$ iff
$mark \cap H_{inp}(tr) = H_{inp}(tr) \wedge mark \cap H_{out}(tr) \backslash H_{inp}(tr) = \emptyset$.

- $LTS_{PN} = \langle N_{PN}, n_{PN0}, W_{PN} \rangle$ |
  - $N_{PN} \subseteq 2^{H_{PN}}$—the set of *nodes*,
  - $n_{PN0} = mark_0$—the *initial node*, root of LTS_{PN},
  - $W_{PN} \subseteq \{(mark_{inp}(tr), tr, mark_{out}(tr)) | tr \in Tr\}$—the set of *LTS transitions.*

7.1.2 Translation of IMDS to PN

IMDS system is translated to PN = (H_{PN}, H_{PN0}, Tr) such that

- $$H_{PN} = M \cup P \tag{7.2}$$
- $$mark_0 = M_{ini} \cup P_{ini} \tag{7.3}$$
- $$Tr : \forall_{\lambda \in \Lambda} \lambda = ((m,p),(m',p')) \text{ iff} \{m,p\} Tr \{m',p'\}; \ \forall_{\lambda \in \Lambda} \lambda = ((m,p),(p')) \text{ iff } \{m,p\} Tr \{p'\}; \tag{7.4}$$

The correspondence between LTS and LTS_{PN}:

1. Correspondence between n_0 and n_{PN0}: $p \in P_{ini}$ iff $p = p_{0i} \in H_{PN0}$, $m \in M_{ini}$ iff $m \in H_{PN0}$ (from (7.3)).
2. Correspondence between N and N_{PN}: $\forall_{T \subseteq H} \exists \ n_{PN} \ n_{PN} \in 2^{H_{PN}}$ (T – IMDS configuration, H_{PN} – set of all IMDS items) (from (7.2)).
3. Correspondence between Λ and Tr:
   (a) In $mark_0$, exactly one place for every server is in $mark_0$ (P_{ini} has exactly one state for every server) and exactly one place for every agent is in $mark_0$ (M_{ini} has exactly one state for every agent) (from (7.3)). Hence, the output places for every enabled PN transition are not marked or they are the same as input places. For every $\lambda \in \Lambda$ the $tr \in Tr$ is defined, therefore for every prepared action λ prepared in $M_{ini} \cup P_{ini}$ the corresponding PN transition tr (having the corresponding input and output) is enabled.
   (b) Firing the PN transition that has a pair of places on input empties these two places and inserts tokens to output places. If the input marking of the PN transition $mark_{inp}(tr)$ has one token for every server and at most one token for every agent, and the output state place p' concerns the same server as the input state place and the output message place m' concerns the same agent

as the input message place (or there is no output message place), then the output marking has still one token for every server and at most one token for every agent. Moreover, the same agents are represented in the output marking or the output marking in smaller by one agent: the agent that is not represented in lacking output message place (from (7.4)). This ensures that the output places of the PN transition are empty or they are the same as input ones.

(c) Having exactly one token for every server and exactly one token for every agent in the initial marking, and the transitions that preserve the number of tokens of the servers and preserve or reduces the latter number by the agent terminating in the PN transition, we get in result that exactly the PN transitions corresponding to prepared IMDS actions are enabled in the input marking equal to the input configuration, and the output marking corresponds to the output configuration.

4. Thus we can conclude that the LTS_{PN} of a PN corresponding to an IMDS system corresponds to the *LTS* of this IMDS system.

7.1.3 Example

The example of the "Two semaphores" system (Fig. 5.1) converted to the Petri net (in ANDL format) is shown below. First, the places with initial marking are defined. Then the transitions follow, with input places marked "−1" and output places "+1". For example, the action $(m1, p1)\Lambda(m2, p2)$ in the server s is modeled as *s_1::[m1-1]&[p1-1]&[m2+1]&[p2+1]:1;*. The final "1" is not important.

```
pn [ two_sem ] {
constants:
places:
A_1_sem_1_wait = 0 ;
A_1_sem_1_signal = 0 ;
A_2_sem_1_wait = 0 ;
A_2_sem_1_signal = 0 ;
sem_1_up = 1 ;
...
transitions:
sem_1_1::[A_1_sem_1_wait-1]&[sem_1_up-1]&[A_1_proc_1_ok_wait+1]&[sem_1_down+1]:1;
sem_1_2::[A_2_sem_1_wait-1]&[sem_1_up-1]&[A_2_proc_2_ok_wait+1]&[sem_1_down+1]:1;
sem_1_3::[A_1_sem_1_signal-1]&[sem_1_down-1]&[A_1_proc_1_ok_sig+1]&[sem_1_up+1]:1;
...
}
```

The report from Charlie (Heiner et al. 2015) contains (among other data) some statistics about elements of the generated Petri net model:Para>

```
input places:
          |12.A_1_proc_1_start:1,
          |15.proc_1_ini:      1,
          |19.A_2_proc_2_start:1,
          |22.proc_2_ini:      1
output places:
          |18.proc_1_stop:     1,
          |25.proc_2_stop:     1
input transitions:
no input transitions
output transitions:
no output transitions
number of places: 29
number of transitions: 20
number of arcs: 78
the net is not connected,
there are 2 components:
C1:
place: A3_r_right, id: 27, orgId: 27
place: A3_r_left, id: 26, orgId: 26
transition: r_19
place: r_res, id: 28, orgId: 28
transition: r_20
C2:
place: proc_2_first, id: 23, orgId: 23
place: A_1_sem_1_signal, id: 1, orgId: 1
transition: proc_1_11
transition: proc_1_10
transition: proc_1_12
place: A_2_sem_2_signal, id: 9, orgId: 9
place: A_2_sem_1_wait, id: 2, orgId: 2
transition: proc_1_13
...
```

The report says that there are two separate components in the net (*C1* is composed of *r* server with *A3* agent and *C2* is composed of servers *sem[1..2], proc[1..2]* and agents *A[1..2]*); no unusable parts of the model (dead code) are reported. In Charlie, communication deadlocks can be found using search for siphons (Schmid and Best 1978; Chu and Xie 1997), see Sect. 7.2.

Figure 7.2 presents graphically the "Two semaphores" system (Fig. 5.1), converted to the Petri net. There are shown three agents *A[1], A[2]* and *A3*

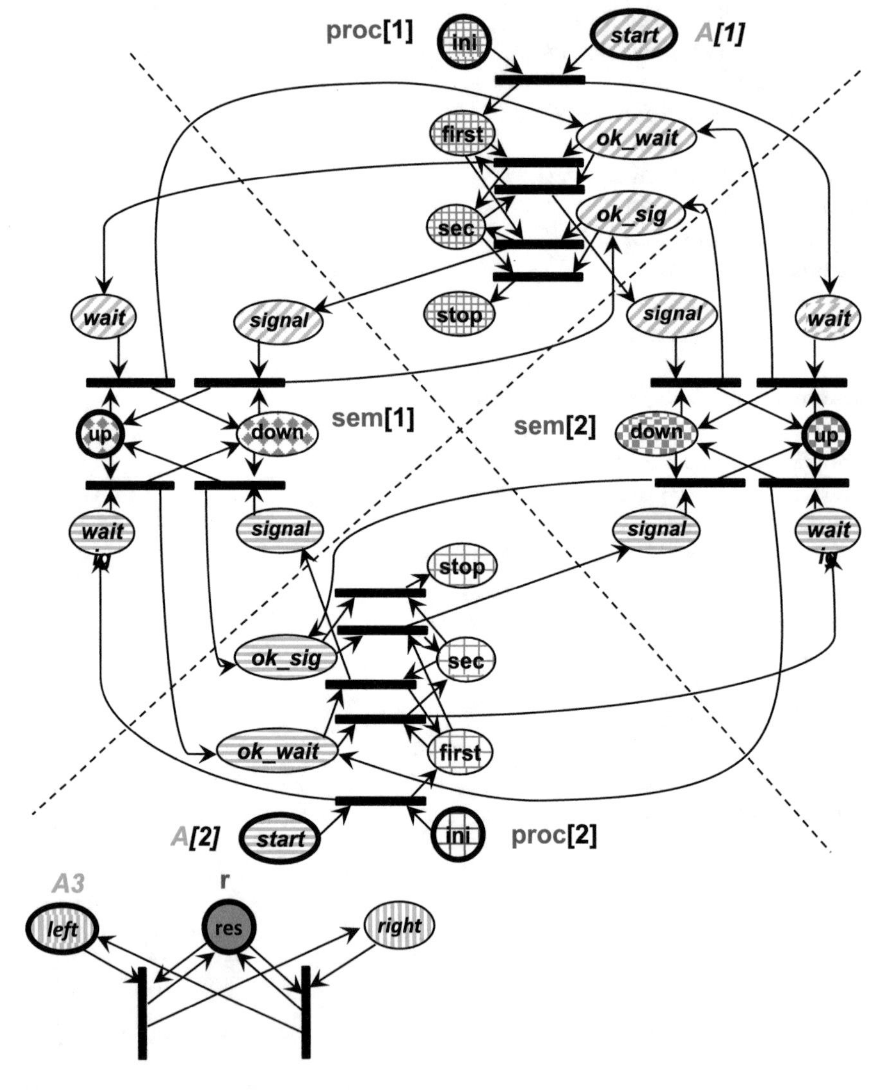

Fig. 7.2 Petri net representation of the "two-semaphores" system (servers *proc[1..2], sem[1..2], r*), agents *A[1..2],A3*)

(names in italics), their servers of origin *proc[1]*, *proc[2]* and *r*, and two semaphores *sem[1], sem[2]*. The initial marking of every server and every agent is surrounded by bold ovals. The states and services concerning individual servers are separated by dashed lines. The places of *proc[1]* and *proc[2]* are filled with a grid pattern (with different density for individual servers), and places of *sem[1]* and *sem[2]* are filled with checkerboard pattern (in diamonds for *sem[1]*). The places of *A[1]* are hatched diagonally and of *A[2]* horizontally). The agent *A3* is hatched vertically and its server *r* has a solid fill.

7.2 Deadlock Detection Using Siphons

A siphon is a Petri subnet that, if loses the tokens, cannot restore them and remains unmarked. If the emptying of a siphon is reachable, it denotes a deadlock (Zuberek 2009).

The advantage of siphon analysis is that multiple total deadlocks can be found in a verified system, while model checking typically finds one deadlocks and then the analysis is stopped. It is because the purpose of model checking is to evaluate temporal formulas, so finding a single configuration providing false result is enough. Conversely, structural analysis of a Petri net allows for identification of all elementary siphons (elementary siphons do not contain other siphons).

Consider a buffer serving two processes, each of which can by a sender or a receiver. In a `neutral` state, an agent can decide whether to `put` or to `get` (lines 16,17). The source code is below:

```
1. #DEFINE N 2
2. #DEFINE K 1

3. server:  buf(agents Aprodcons[N];servers Sprodcons[N]),
4. services {put, get},
5. states    {elem0,elem[K]},
6. actions  {
7. <i=1..N>{Aprodcons[i].buf.put, buf.elem0}->
           {Aprodcons[i].Sprodcons[i].ok_put, buf.elem[1]},
8. <i=1..N><j=1..K-1>{Aprodcons[i].buf.put, buf.elem[j]}->
           {Aprodcons[i].Sprodcons[i].ok_put, buf.elem[j+1]},
9. <i=1..N><j=2..K>{Aprodcons[i].buf.get, buf.elem[j]}->
           {Aprodcons[i].Sprodcons[i].ok_get, buf.elem[j-1]},
10. <i=1..N>{Aprodcons[i].buf.get, buf.elem[1]}->
           {Aprodcons[i].Sprodcons[i].ok_get, buf.elem0}
11. }
```

```
12. server:   Sprodcons(agents Aprodcons;servers buf),
13. services {doSth,ok_put,ok_get}
14. states   {neutral,prod,cons}
15. actions  {
16. {Aprodcons.Sprodcons.doSth, Sprodcons.neutral} ->
             {Aprodcons.buf.put, Sprodcons.prod}
17. {Aprodcons.Sprodcons.doSth, Sprodcons.neutral}->
             {Aprodcons.buf.get, Sprodcons.cons}
18. {Aprodcons.Sprodcons.ok_put, Sprodcons.prod}->
             {Aprodcons.Sprodcons.doSth, Sprodcons.neutral}
19. {Aprodcons.Sprodcons.ok_get, Sprodcons.cons}->
             {Aprodcons.Sprodcons.doSth, Sprodcons.neutral}
20. }

21. servers  buf,Sprodcons[N];
22. agents   Aprodcons[N];

23. init      ->{
24. <j=1..N>Sprodcons[j](Aprodcons[j],buf).neutral,
25.         buf(Aprodcons[1..N],Sprodcons[1..N]).elem0,

26. <j=1..N>Aprodcons[j].Sprodcons[j].doSth,
27. }.
```

The example was converted to ANDL format, where the place names are compound of agents with their messages or servers with their states. For example, the message (*Aprodcons[1], buf, put*) is converted to `Aprodcons_1_buf_put`), and the state (*buf, elem[1]*) is converted to `buf_elem_1`.

The system was subject of siphon analysis in Charlie. There are 49 elementary siphons, but we show how abstraction classes can be built over them. Below is a list of elementary siphons (called minimal in Charlie):

```
minimal siphon ( place ) =
1        |0.Aprodcons_1_buf_put    :1,
         |2.Aprodcons_2_buf_put    :1,
         |5.buf_elem_1        :1,
         |6.Aprodcons_1_Sprodcons_1_doSth       :1,
         |8.Aprodcons_1_Sprodcons_1_ok_get      :1,
         |10.Sprodcons_1_prod       :1,
         |14.Aprodcons_2_Sprodcons_2_ok_get    :1,
```

```
            |15.Sprodcons_2_neutral  :1,
            |16.Sprodcons_2_prod       :1
2           |1.Aprodcons_1_buf_get    :1,
            |3.Aprodcons_2_buf_get    :1,
            |4.buf_elem0            :1,
            |6.Aprodcons_1_Sprodcons_1_doSth          :1,
            |7.Aprodcons_1_Sprodcons_1_ok_put         :1,
            |11.Sprodcons_1_cons       :1,
            |13.Aprodcons_2_Sprodcons_2_ok_put        :1,
            |14.Aprodcons_2_Sprodcons_2_ok_get        :1,
            |15.Sprodcons_2_neutral  :1
            ... (etc)
```

The system has two obvious deadlocks: getting from an empty buffer by both agents and putting to a full buffer by both agents. There should be siphons that can be emptied, representing the two deadlocks. There are 49 elementary siphons found. Every empty siphon can be checked for reachability using the model checking technique (in Charlie's output, the set of places is listed that should be emptied, and every emptied place represents the lack of message or state in the configuration), using the CTL formula **AG** ($\neg\ \varphi$) (or LTL formula $\Box$ ($\neg\varphi$)), where φ is the empty siphon. In the example above, the Uppaal formula (Uppaal (Behrmann et al. 2006) is one of external model checkers used in Dedan, it is described in Chap. 9) for the siphon number 1 is (services has the encoding 1-`put`, 2-`get`, 1-*`doSth`*, 2-*`ok_put`*, 3-*`ok_get`*):

```
A[] !(Aprodcons_1_buf!=1 & Aprodcons_2_buf!=1 & !buf_.elem_1 &
Aprodcons_1_Sprodcons_1!=1 & Aprodcons_1_Sprodcons_1!=3 &
!Sprodcons_1.prod & Aprodcons_2_Sprodcons_2!=3 & !Sprodcons_2.neutral &
!Sprodcons_2.prod)
```

Verification gives the result *false*, which means that the siphon can be emptied. Uppaal generates a counterexample, in which both agents perform `get` on an empty buffer (state *`elem0`*). Many other siphons can be emptied by two `get` operations on an empty buffer. All these situations constitute a single deadlock (but the counterexamples can differ in the order of issuing `get` by the two agents). Therefore, each siphon can be easily checked for emptying (there are some siphons that are not emptied in the example, for instance *`(buf_elem0,buf_elem_1)`*). Because some siphons are equivalent (they have the same or equivalent counterexamples, differing in the order or operations), abstraction classes should be made in the set of siphons. The criterion for the allocation of siphons to the abstraction classes is the configuration that finishes the counterexample (it is equivalent to the corresponding marking in the Petri net). In this example, there are two abstraction classes representing the two possible deadlocks, they are identified by the configurations (triples are messages, pairs are servers' states):

```
1. {(Aprodcons[1],Sprodcons[1],get),
   (Aprodcons[2],Sprodcons[2],get), (Sprodcons[1],cons),
   (Sprodcons[2],cons), (buf,elem0)}
2. {(Aprodcons[1],Sprodcons[1],put),
   (Aprodcons[2],Sprodcons[2],put), (Sprodcons[1],prod),
   (Sprodcons[2],prod), (buf,elem[1])}
```

For small systems, for which the LTS is prepared internally in Dedan, the reachability of siphon emptying can be checked directly over the LTS. For example, for siphon number 1 the reachability of siphon emptying can be verified by the inspection of the LTS and finding a configuration in which:

- pending message of the agent *Aprodcons[1]* is not *put* at *buf*, or the agent is terminated (no pending message of *Aprodcons[1]*),
- pending message of the agent *Aprodcons[2]* is not *put* at *buf*,
- current state of the server *buf* is not *elem_1*,
- pending message of the agent *Aprodcons[1]* is not *doSth* at *Sprodcons[1]*,
- pending message of the agent *Aprodcons[1]* is not *ok_get* at *Sprodcons[1]*,
- current state of the server *Sprodcons[1]* is not *prod*,
- pending message of the agent *Aprodcons[2]* is not *ok_get* at *Sprodcons[2]*,
- current state of the server *Sprodcons[2]* is not *neutral*.

Because a siphon may be emptied in various ways, it can lead to more than one deadlock. Model checking identifies one example of emptying the siphon in a single run. Therefore, our procedure does not guarantee the identification of all deadlocks in a single run, one deadlock is found per a reachable empty siphon. Nevertheless, the ability to identify multiple deadlocks (one for every abstraction class of emptied siphons) in a single procedure is an advantage. In addition, the described procedure liberates from constraining siphon-based deadlock detection only to purely cyclic systems.

7.3 Problem with Not Purely Cyclic Systems

The example of bounded buffer is a purely cyclic system. If a system is not cyclic, it may contain a *leader*—the initializing subnet of the system before it starts to operate cyclically. Such a leader obviously contains a siphon—an emptied subnet which does not restore tokens (because it is not needed). Such a system of "Two semaphores" is described in Sect. 5.1. The siphon is found for every place that implements the initial state of the server or the initial message of the agent (siphons number 4, 5, 7, 8, 9):

```
minimal siphon ( place ) =
1          |10.sem_2_up          :1,
           |11.sem_2_down        :1
2          |26.A3_r_left         :1,
           |27.A3_r_right        :1
3          |1.A_1_sem_1_signal         :1,
           |3.A_2_sem_1_signal         :1,
           |4.sem_1_up :1,
           |7.A_1_sem_2_signal        :1,
           |9.A_2_sem_2_signal         :1,
           |10.sem_2_up          :1,
           |13.A_1_proc_1_ok_wait    :1,
           |14.A_1_proc_1_ok_sig     :1,
           |20.A_2_proc_2_ok_wait   :1,
           |21.A_2_proc_2_ok_sig     :1
4          |15.proc_1_ini       :1
5          |22.proc_2_ini       :1
6          |4.sem_1_up :1,
           |5.sem_1_down         :1
7          |19.A_2_proc_2_start       :1
8          |12.A_1_proc_1_start      :1
9          |28.r_res      :1
```

For the siphon number 7, formula `A[] !(A_2_proc_2! = 1)` evaluates to *false*, suggesting a deadlock, but it is the initial agent message (*A[2]*, *proc*, *start*), included in the leader. Figure 7.3 shows two types of initial siphon that can be emptied, the leader siphon is depicted on the left (Fig. 7.3a) and another type of siphon on the right (Fig. 7.3b). In the figure "Ending Strongly Connected Subgraph" is a cycle from which there is no escape. Alternatively, the system can have a terminating trailer part (reverse counterpart of the leader), which can contain the emptyable siphon. The pictures are schematic, showing the overall shape of the system. Examples of not purely cyclic systems, known from the literature, are:

- "linear" systems (such as "two semaphores" described in Sect. 5.1: users issue *wait* to two semaphores, then they issue *signal* and terminate),
- a system with "leader" (initial part) and a main loop, sometimes called "lasso-shaped" (Latvala et al. 2004),
- terminating system with a main loop, for example WF-net system (van der Aalst 2000),

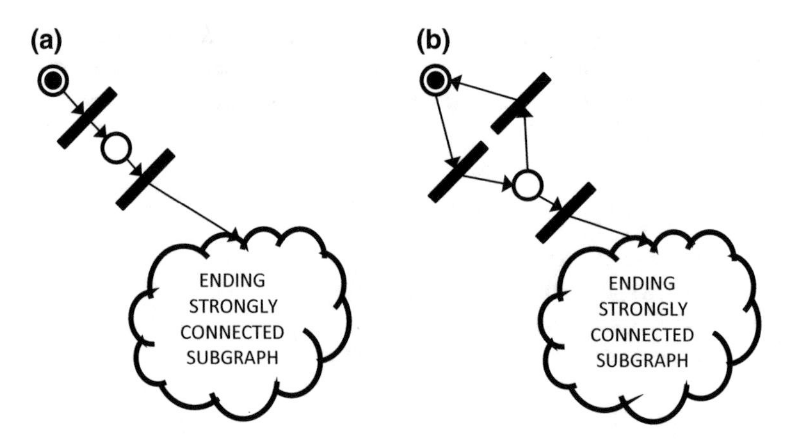

Fig. 7.3 Siphons: **a** a leader siphon; **b** an initial siphon not being a leader siphon

- similar to lasso-shaped, but with an initial loop (Fig. 7.3b),
- a system with cyclical part, and an escape from it to the terminating trailer.

In some cases, for example in WF-nets, the system can be easily converted to a cyclic one, connecting initial and terminating places. However, in the analysis of distributed systems, especially those following the IoT paradigm, a system can have multiple leaders (for every node) and multiple terminating places, where the nodes reach their goals. An example is Automatic Vehicle Guidance System presented in Sect. 5.4.

To solve such problems, a procedure was made to verify if the siphon denotes a deadlock:

1. For every siphon found by a Petri net analysis tool (like Charlie), a formula φ is constructed, as described in Sect. 7.2. The siphon contains a set of places being states and messages.
2. The possibility of emptying the siphon is tested, using the formula **AG** ($\neg$ φ).
3. If the result is false, the siphon can me emptied, and for each server and agent its continuation, the formula is verified: **AG**($\varphi \Rightarrow$ **EF** $\neg$ (φ/restricted to the process)). This identifies the processes involved in the deadlock.

After this procedure, the following cases can be concluded based on the result of the evaluation of the formula in p.3:

- the formula is *true*—the process is not in deadlock,
- the formula is *true* for all processes—the siphon is not a deadlock,
- the formula is *false* for all processes—total deadlock,
- the formula is *false* for selected processes—partial deadlock, agents/servers with *false* result participate.

In "Two semaphores" example, the siphons listed above denote:

- 1, 2, 6, 9: not emptyable siphons – no deadlock,
- 3: deadlock siphon of `sem_1` and `sem_2` in the server view, and of `proc_1` and `proc_2` in the agent view,
- 4, 5, 7, 8: emptyable leader siphons – no deadlock.

7.4 Translation from Petri Net to IMDS

The IMDS formalism is sometimes regarded as exotic to people who are used to apply other formalisms. That is why we describe the invention of IMDS step by step, adding the natural features of distributed systems. For this purpose, the well-known formalism of Petri nets is used. We are making the following assumptions about Petri net:

- Places capacity equal to 1.
- Interleaving semantics (one transition is fired at a time).

7.4.1 Servers

Our requirement is that we should do not arbitrarily construct places. If we give a new attribute (with a set of values) to the set of places, then we can split the places accordingly, assigning each value of the new attribute to one of the new places.

We see the system as a node—a single place (Fig. 7.4a). The system runs—it performs some actions (Fig. 7.4b). If the system is distributed, we see its activity as messages exchanged inside. We model the system as the entity place and the message place (but without identifying individual messages), to distinguish the system structure from its behavior (Fig. 7.4c). This shows the assumptions imposed on the model:

1. The action is caused by a message.
2. The action causes the next message.

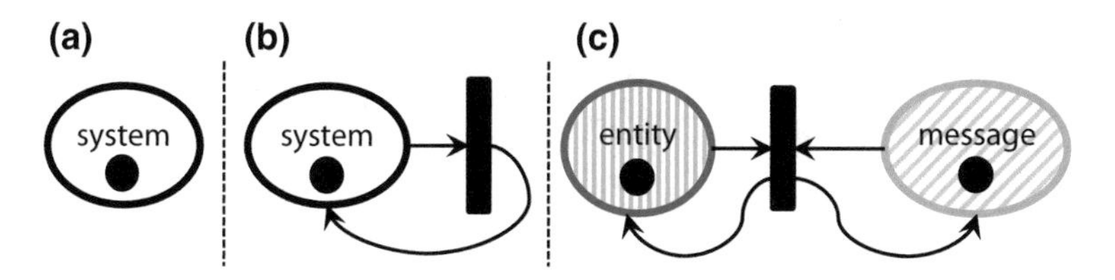

Fig. 7.4 The model: **a** a system; **b** working system; **c** working system with messages

In addition to the colors used in this monograph for servers and their states (red), agents and their messages (green), we also use different hatching (vertical/ horizontal vs. diagonal) and font (regular vs. *italics*).

Having extracted messages, we see that the system accepts the message and causes the next message to appear. To run such system, we need tokens in the entity place and in the message place. The latter can be interpreted as the initial message starting the calculation run and then being the carrier of the calculation.

From now on, we require that each transition has two input arcs: one from an entity place (after proper splitting, below we call them server places) and the other from a message place. We call them input places of a transition.

At the same time, we require for every transition (with the exception of termination, described later) to have two output arcs: one to the entity place and the other to the message place. We call them output places.

As the system is distributed, we distinguish its components—the servers, replacing the single entity. Messages are sent just between these servers, and each message has a target server (Fig. 7.5a). In the figure, individual server places and message places differ in color and hatch density. The server place and the message sent to this server have the same hatching direction.

We can see that the server can send a message to itself or to another server. In the system in Fig. 7.5a, sending to the other server is blocked due to the occupancy of each message place by the token. The only option is to simultaneously fire the two transitions sending messages to another server (*S* to *T* and *T* to *S*), which is prohibited due to interleaving semantics. Therefore, in the example we give the initial token only to one of the messages (Fig. 7.5b), but this is not a general requirement and we will be able to give up on this after the next splitting of places.

The transition models the activity of the server, therefore we require that the transition has input arcs concerning the same server (for example, arcs from "Server

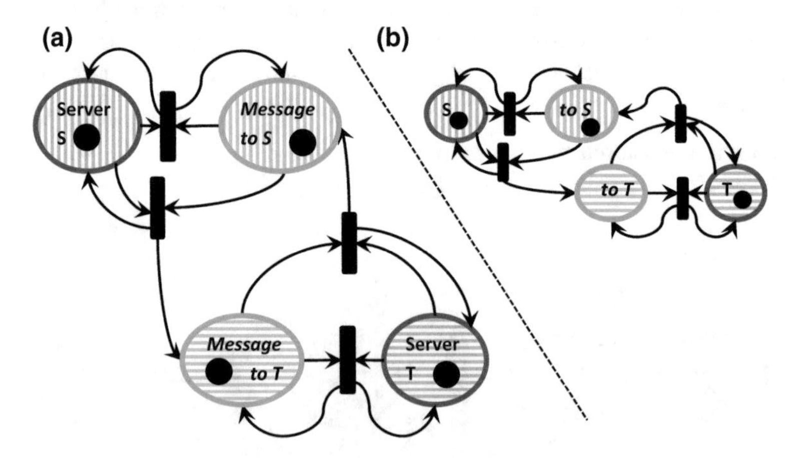

Fig. 7.5 **a** System consisting of two servers; **b** modified—initial token retrieved from a message to *T*

S" place and "Message to *S*" place). One of the output arcs must lead to the same server as the input arc from the server place, to preserve the server activity. The other output arc can lead to any server, because it models sending a message.

At this point, we see the next assumption imposed on the distributed system model:

3. The server can send a message to itself.

An additional feature is a *restriction*: the action cannot cause a set of messages directed to the servers (only one message is caused by the action). This is of course our own assumption and not a general requirement.

7.4.2 Agents

To introduce the next feature, we want to have three servers, like in Fig. 7.6a. The system is similar to that in Fig. 7.5, with one server added. No server sends messages to itself, which happens in Fig. 7.5. The servers *T* and *U* can send messages to the server *S* (and accept response messages), but servers *T* and *U* do not communicate with each other directly. Note that although both *T* and *U* servers can send messages to the server *S*, the messages cannot be pending at *S* simultaneously because the capacity of the place "Message to *S*" is 1. Each server place has a renewable token which means that the server is always ready to accept a message.

Now we may say that some distributed computations are performed in the system. There may be several such computations. Even a single server can start more than one computation. We introduce a new entity called agent, which represents a distributed computation. The agent defines a sequence of actions to be performed on the servers in the system, caused by messages attributed to this agent.

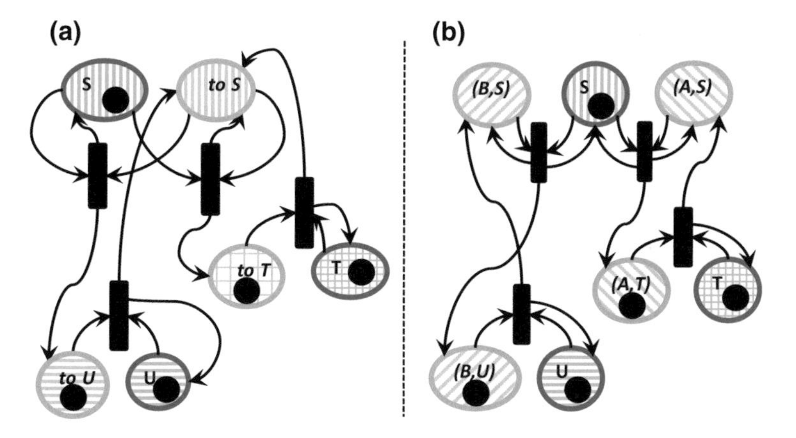

Fig. 7.6 **a** A system consisting of three servers *S*, *T* and *U*; **b** agents *A* and *B* introduced

In this way, we change the message place attributes: they are pairs (*server*, *agent*), and the meaning of the pair is "message sent to the server in the context of the agent". At the same time we withdraw the "message to server" attribute from the message places. This is illustrated in Fig. 7.6b, where agents *A* and *B* are introduced. Each server has a different hatch (horizontal, vertical or grill) and each agent has its own hatch (diagonal). The same hatch scheme is used in the following figures.

Now, two messages (*A*, *S*) and (*B*, *S*) can be pending at the server *S* at the same time. Of course only one of the actions caused by these two messages can be fired, because of interleaving semantics.

The transition has two server places of the same server on input and on output. The second input place is the agent's message. To obtain a continuous computation, the output message place must have the same agent attribute as the input message place.

To start a computation, exactly one of the places of the agent has an initial token. By construction, always exactly one of agent places contains a token, because the transition transfers a token from its input message place to its output message place (except the termination, described later).

During the system run, some situations may prohibit certain actions while others are enabled. For example, sending an element to a full buffer is forbidden, while sending it to an empty buffer is possible. To distinguish such situations, we introduce a set of states attributed to every server. The server state (full or empty buffer in the example) is defined by the value of all its internal variables. In this way, we change the server place attributes: they are pairs (*server*, *value*), and the meaning of a pair is "server state value". At the same time we withdraw the "server" attribute from server places. This is illustrated in Fig. 7.7a, where values *V* and

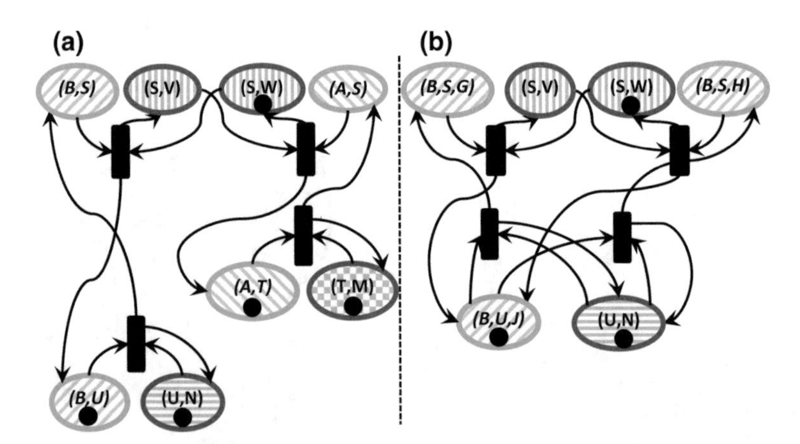

Fig. 7.7 **a** Server states introduced to the system (3 servers *S*, *T*, *U*, 2 agents *A*, *B*), server *S* has two states with values *V* and *W*; **b** services distinguishing server calls (2 servers *S*, *U*, 1 agent *B*), server *S* offers two services *G* and *H*

W are introduced into the server *S*. The server *T* has only one state with value *M* and the server *U* with value *N*. For clarity, the states of a given server and the messages of a given agent are shaded in uniform way.

Each server has an initial token in exactly one of its places. By construction, always exactly one of the server's places contains a token (it is the initial value of the server state).

Until now, the agent can send a message to the server. But in distributed systems, the server can offer several different operations, for example, granting the resource and releasing it. Such operations are called server services. Even a single agent can call different services, depending on the progress of the calculation. Such a system is illustrated in Fig. 7.7b, where agent *B* causes the server *S* to switch between two states *V* and *W*, by invoking the services *H* and *G*. The server *U* offers the only service *J*. The service is the third attribute of the message place. The two actions executed on the server *U* invoke the services *G* and *H* of the server *S*.

7.4.3 Example

An example of a system consisting of a 1-element buffer and two producers/consumers (switching nondetrministically between producing and consuming) is presented in Fig. 7.8. The buffer and the two users have their own servers: `S1`, `S2` and `buf`. Each user is equipped with its own agent (`A1`/`A2`) which produces/consumes elements and puts them to/gets from the buffer. In each user's server, the state `n` is neutral, `prod` is producing and `cons` is consuming. The services invoked at the buffer are `put` and `get`. The buffer sends responses `ok_put` and `ok_get`. After receiving a response, the agent issues a message `doSth` to its own server and then chooses producing or consuming, in a nondeterministic way.

7.4.4 Termination

In some systems the computations can finish. For this purpose, a special type of transition is defined: no output message place is present. Figure 7.1 shows a regular action and a terminating action.

7.4.5 Processes

Now it's time to define processes in the system. Note that we defined a structure of the system, without the concept of processes. In this way, the processes are

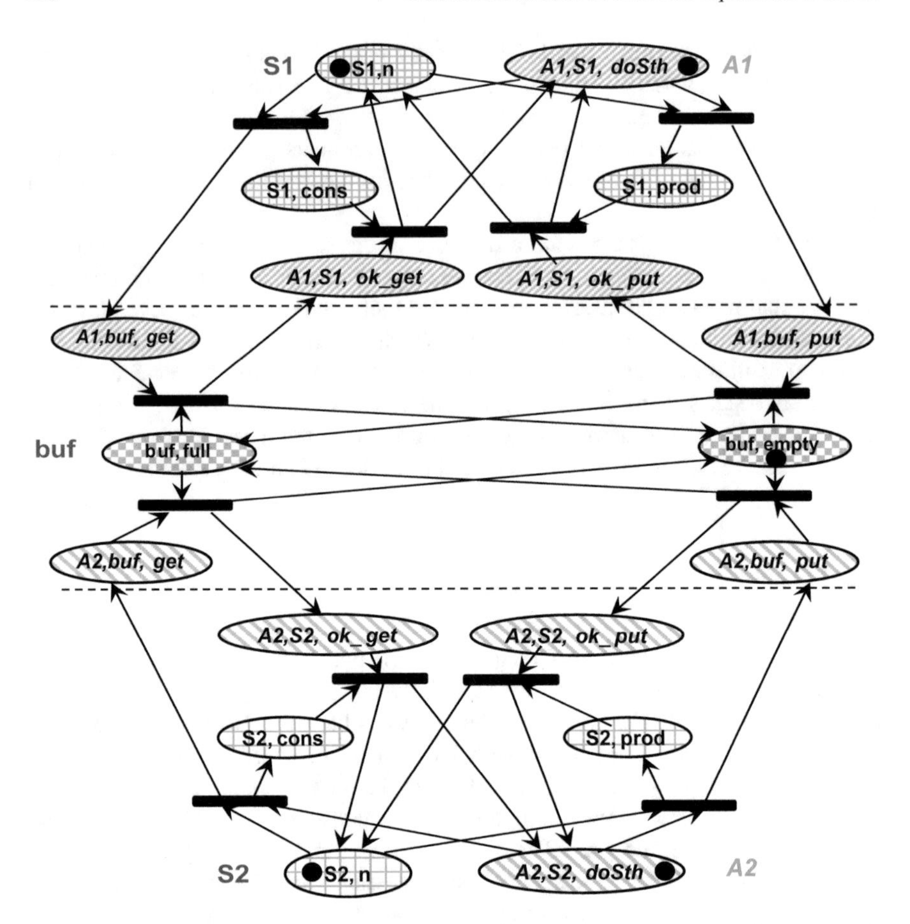

Fig. 7.8 Petri net representation of the "bounded buffer" system: servers *S*1, *S*2, *buf*, agents *A*1, *A*2

extracted from the system, rather than the system is constructed over processes. This approach is on purpose, because processes can be defined in various ways. We focus on asynchronous processes.

The assumption is that we build processes over transitions. The process is a sequence of transitions. It is a loop in a cycling system, or a finite sequence in a terminating system. Of course, mixed structures are possible, for example "lasso-shaped" system, where initializing sequence is followed by an infinite loop, or a system terminating with a trailer part.

In Petri net often processes are defined over transitions. Usually, places are divided into two subclasses: process carriers and resources. The place connecting two transitions in a sequence is the process carrier. We say that the connecting place "threads" the process. The other places model the resources, acquired in arcs leading to transitions and released in arcs leading from transitions.

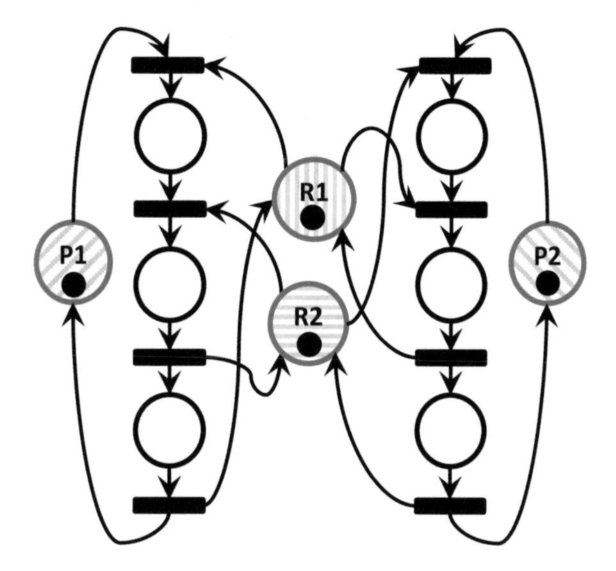

Fig. 7.9 Petri net modeling two processes (*P*1 and *P*2, violet) using two resources (*R*1 and *R*2, blue)

In our model, each transition has two input and two output places (with the exception of agent-terminating actions). This looks similar to the process/resource model above, but messages occur instead of resources. It is illustrated in Fig. 7.9, based on (Zuberek 1999). Process places have different diagonal hatching and resource places have horizontal and vertical hatching. A similar approach is described in López-Grao and Colom (2010), where process arcs are shown as solid arrows while resource allocation arcs are dashed. Acquiring the resource and releasing it are separated from each other, therefore the transitions have 1 input place/2 output places or vice versa, but this difference does not change much.

The main difference is that agent messages appear in sequence, traveling between servers (possibly multiple servers), while the resource place loses the token just for the time the resource is used, and then receives the token back. Note that there are papers in which a duality other than ours is described, in Wu and Zhou (2010) two Petri net types are considered: process-oriented and resource-oriented. However, those are not two aspects or "views" of a uniform net. The dichotomy between the process view and the resource view is modeled using two different Petri net structures that serve to model the views. In our approach, these are the two definitions of processes in a uniform net, rather than two separate nets. Note that in our model, the resources are not modeled directly, because in a distributed environment, resource allocation is not performed as a single operation. Resources are distributed over the network, therefore the minimum activity to obtain a resource is to send a request, wait for grant and obtain a grant confirmation. Therefore, everything lies in communication.

In our model, we can choose which of the places "threads" the process: state place or message place. The answer is: whatever you want. Each of the input places can be chosen as a process carrier. If we choose state places, we get the server process. It consists of actions threaded by state places. However, we may choose

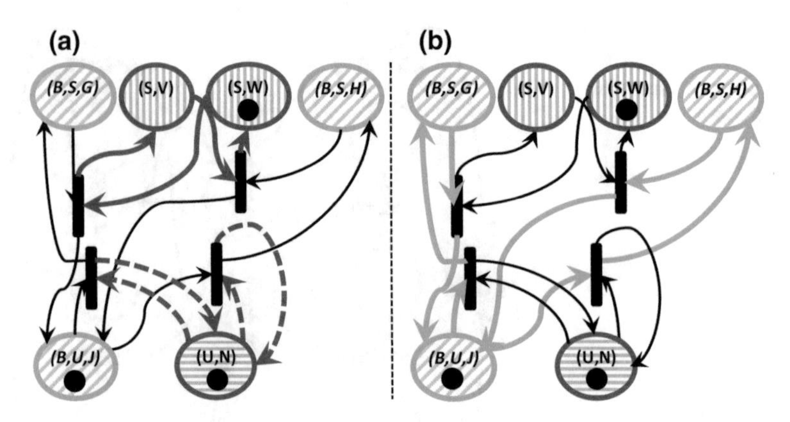

Fig. 7.10 **a** Two server processes *S* and *U*; **b** one agent process *B*

message places gluing a process, resulting in the agent process. In Fig. 7.10, we have the system from Fig. 7.6, with two server processes highlighted (Fig. 7.10a) and one agent process highlighted (Fig. 7.10b). The arcs between the process transition and the places gluing the process are thickened and colored with appropriate color (red for servers, green for agents). In addition, the arcs of the server process *U* are distinguished by dashing.

In the system, exactly one of the places of the server has a token. Similarly, exactly one of the places of an agent has a token (or at most one place, if the agent terminates). Therefore, we can change the requirement 1 (the capacity of places is 1), restricting it to the initial places of servers and agents. In other places, the requirement is achieved by construction.

Another problem arises: if we choose the places of given kind (state places or message places) as a process carrier, what about the other places? The answer is simple: the remaining places serve as means of communication. Now we can give two projection names: server view and agent view. In the server view, state places are the carriers of server processes while message places are the means of communication. In the agent view, message places are the carriers of agent processes while state places are the means of communication. The behavior is uniform, because it is a uniform system. Behavior is modeled by the reachability graph of the markings.

7.4.6 Formal Requirements for Petri Net

Now we show, how to formally build the Petri network to fit our assumptions.

1. Interleaving semantics (one transition is fired at a time).
2. Each transition has two input places: server state (s, v) and message (a', s', r'). The server component of both places must match $(s = s')$ (Fig. 7.11a).

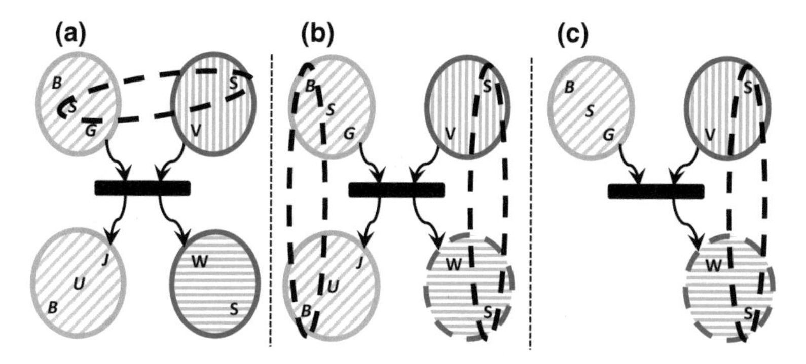

Fig. 7.11 Required matching of server ID and agent ID in an action: **a** server ID (*S*) matches in input places; **b** the same agent ID (*B*) in input and output message places, the same server ID (*S*) in input and output state places, **c** agent-terminating transition—the same server ID (*S*) in input and output state places

3. Each regular transition has two output places: the new server state (s, v'') and the next agent message (a, s''', r'''). Input and output state places have the same server attribute, while input and output agent places have the same agent attribute (Fig. 7.11b).
4. The agent-terminating transition has only one output place: the new server state (s, v''). The input and output state places have the same server attribute (Fig. 7.11c).

We must supplement the principles with initial marking:

5. Among the places of any server's states, exactly one has an initial token. It is the initial state of the server (Fig. 7.8, for example the state place ($S1$, n) of the server $S1$).
6. Among the places of any agent's messages, exactly one has an initial token. It is the initial message of the agent (Fig. 7.8, for example the message place ($A1$, $S2$, *doSth*) of the agent $A1$).

7.4.7 Communication Duality, Locality, Autonomy and Asynchrony

The proposed construction of the Petri net exposes the real properties of distributed systems:

- *Communication duality*: a token moving through places of the server's states is the carrier of the server process, while the tokens in matching message places are means of communication (Fig. 7.12a, server process *S*, *T*, *U*). In the dual view, a token "traveling" through the agent's message places is the carrier of the

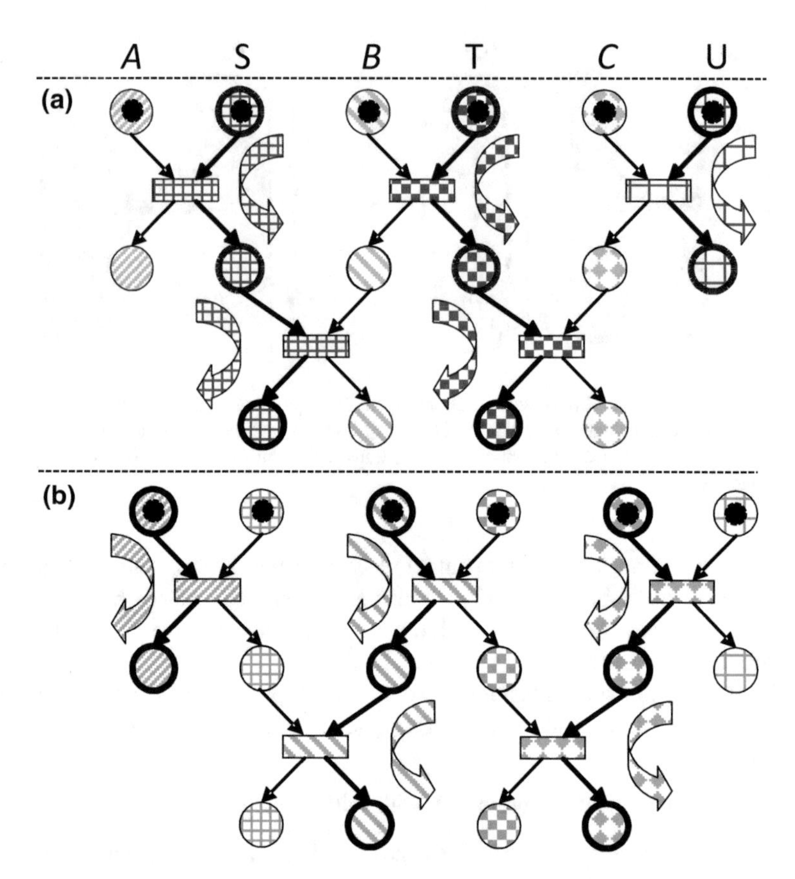

Fig. 7.12 Dual views of a distributed system: **a** server view; **b** agent view; servers *S*, *T* and *U*, agents *A*, *B* and *C*

agent's process, while the tokens in matching state places are means of communication (Fig. 7.12b, agent processes *A*, *B*, *C*). In the figure, transitions have server colors in the server view and agent colors in the agent view. Transitions "belong" to the server processes in the server view and to the agent processes in the agent view. The semicircular arrows in the figure show the flow of processes: (a) server processes following the servers states (bold), (b) agent processes following the messages (bold), traveling between the servers.

- *Locality*: the server transition is made only based on its current state and messages pending at this server. The only manner to influence the behavior of the server is to send a message to it. This is illustrated in Fig. 7.13a: three messages are pending at the server *S*, two of them match the current state of the server and can fire the transitions (of agents *A* and *B*), the third message does not match (of agent *C*), and the message of the agent *D* matches but is not pending.

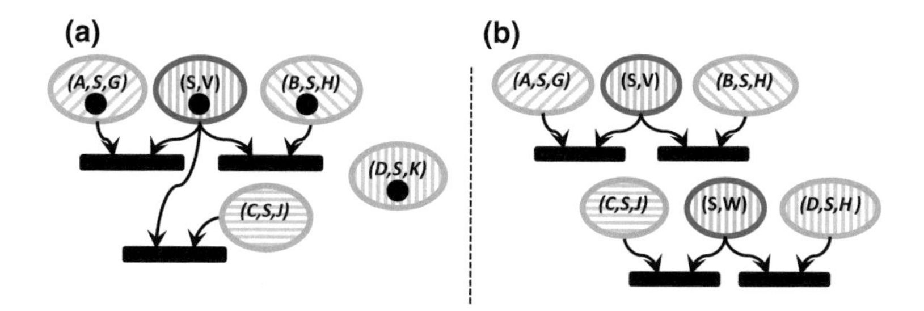

Fig. 7.13 **a** Locality of transitions; **b** autonomy of servers

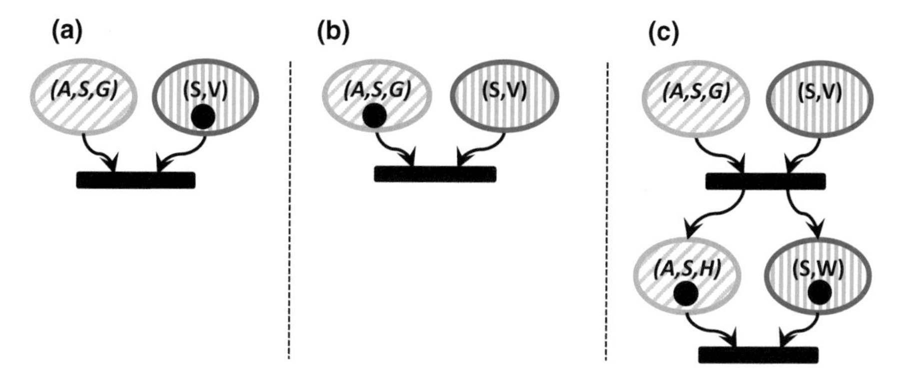

Fig. 7.14 Asynchrony of transitions: **a** state waits for message; **b** message waits for state; **c** message sent by a server to itself – state and message appear synchronously

- *Autonomy*: only the server decides autonomously which messages would be accepted, it is controlled by matching between states and messages. In Fig. 7.13b, state *V* matches messages of agents *A* and *B* invoking services *G* and *H*, while state *W* matches messages of agents *C* and *D* invoking services *J* and *H*.
- *Asynchrony of actions*: usually the current state waits for a matching message (Fig. 7.14a) or the pending message waits for a matching state (Fig. 7.14b), the two tokens appear synchronously only when a message is sent to the sending server itself (Fig. 7.14c: the next state and the next message appear synchronously on the output of the transition, but this case appears entirely inside a process: of the server or of the agent).
- *Asynchrony of communication*: the message is sent via a unidirectional channel, without response. The answer can come through a separate unidirectional channel.

7.4.8 Semantics

All possible system behaviors are included in its graph of reachable markings. The graph for the system shown in Fig. 7.15a is presented in Fig. 7.15b. The form of transitions (dashed/dotted) is adopted from agent migration arcs in Fig. 7.15a. The terminal markings in the lower row denote deadlocks: no message matches any server's state, but the message is still pending at the server *S*. The service of the server *S* is invoked but it will never be executed because the server cannot change its state.

The graph is identical to the Labeled Transition System of IMDS system. The markings in Fig. 7.14b are just current states and pending messages. And they are just the configurations of the system. Likewise, the transitions in Fig. 7.15b correspond to the actions of IMDS. Therefore, the LTS of IMDS system is identical to the graph of reachable markings of the Petri net.

Thus, for designers who are accustomed to model their systems in Petri nets, and who find IMDS formalism to be exotic, the presented scheme should be attractive. The design in Petri net using the scheme presented in Fig. 7.1:

- allows expressing locality, autonomy and asynchrony of distributed systems,
- preserves communication duality, emphasizing the specific features of the system,
- provides simple extraction of process from the model (server processes/agent processes),
- allows finding deadlocks by identifying siphons or using model checking (after conversion to IMDS, which is obvious),
- allows checking the system for inevitable termination, if it is desired (again in the IMDS form).

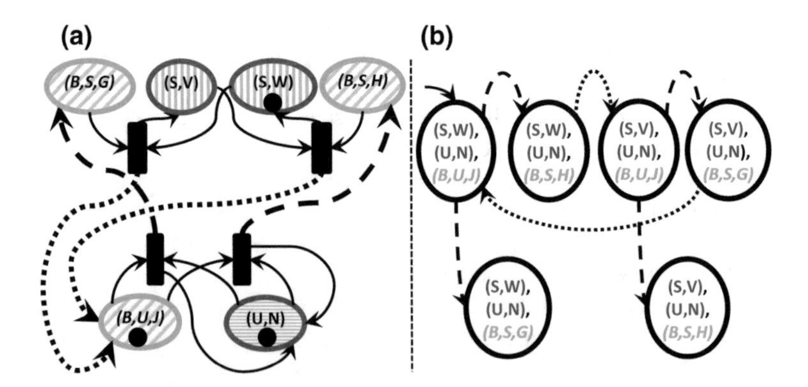

Fig. 7.15 **a** System from Fig. 7.10; transition in the servers shown near their states; messages sent from *U* to *S* shown as dashed bold arrows, responses (messages as well) showed as dotted bold arrows; **b** reachable markings graph; a node includes the names of marked places

References

Behrmann, G., David, A., & Larsen, K. G. (2006). *A tutorial on Uppaal 4.0*. Aalborg, Denmark. URL: http://www.it.uu.se/research/group/darts/papers/texts/new-tutorial.pdf.

Charlie. (n.d.). *Charlie Petri net analyzer*. URL: http://www-dssz.informatik.tu-cottbus.de/DSSZ/Software/Charlie.

Chu, F., & Xie, X.-L. (1997). Deadlock analysis of Petri nets using siphons and mathematical programming. *IEEE Transactions on Robotics and Automation, 13*(6), 793–804. https://doi.org/10.1109/70.650158.

Daszczuk, W. B., & Zuberek, W. M. (2018). Deadlock detection in distributed systems using the IMDS formalism and petri nets. In W. Zamojski, J. Mazurkiewicz, J. Sugier, T. Walkowiak, & J. Kacprzyk (Eds.), *12th International Conference on Dependability and Complex Systems, DepCoS-RELCOMEX 2017, Brunów, Poland.* AISC (Vol. 582, pp. 118–130), July 2–6, 2017. Cham, Switzerland: Springer. https://doi.org/10.1007/978-3-319-59415-6_12.

Heiner, M., Schwarick, M., & Wegener, J.-T. (2015). Charlie—An extensible Petri net analysis tool. In *36th International Conference, PETRI NETS 2015, Brussels, Belgium* (pp. 200–211), June 21–26, 2015. Cham, Switzerland: Springer. https://doi.org/10.1007/978-3-319-19488-2_10.

Jensen, K., & Kristensen, L. M. (2009). *Coloured petri nets*. Berlin, Heidelberg: Springer. https://doi.org/10.1007/b95112.

Latvala, T., Biere, A., Heljanko, K., Junttila, T. (2004). Simple bounded LTL model checking. In A. J. Hu & A. K. Martin (Eds.), *International Conference on Formal Methods in Computer-Aided Design, Austin, TX. LNCS 3312* (pp. 186–200), November 15–17, 2004. Berlin, Heidelberg: Springer. https://doi.org/10.1007/978-3-540-30494-4_14.

López-Grao, J.-P., & Colom, J.-M. (2010). The resource allocation problem in software applications: A Petri net perspective. In S. Donatelli (Ed.), *RAPNeC-2010 Recent Advances in Petri Nets and Concurrency. Edition: Volume 827 of CEUR Workshop Proceedings* (pp. 219–233). CEUR. URL: http://ceur-ws.org/Vol-827/19_Juan-PabloLopez-Grao_article.pdf.

Schmid, H. A., & Best, E. (1978). A step towards a solution of the liveness problem in Petri nets. Newcastle upon Tyne. URL: http://www.cs.ncl.ac.uk/publications/trs/papers/114.pdf.

Schwarick M. (2013). Marcie—An analysis tool for generalized stochastic petri nets. *Cottbus*. URL: http://www-dssz.informatik.tu-cottbus.de/publications/papers/MARCIE_manual_oct_2013.pdf.

van der Aalst, W. M. P. (2000). Workflow verification: finding control-flow errors using petri-net-based techniques. In W. van der Aalst, J. Desel, & A. Oberweis (Eds.), *Business Process Management. LNCS, Vol.1806* (pp. 161–183). Berlin, Heidelberg: Springer. https://doi.org/10.1007/3-540-45594-9_11.

Wu, N., & Zhou, M. (2010). Process vs. resource-oriented Petri net modeling of automated manufacturing systems. *Asian Journal of Control, 12*(3), 267–280. https://doi.org/10.1002/asjc.184.

Zuberek, W. M. (1999). Petri net models of process synchronization mechanisms. In *SMC'99 IEEE International Conference on Systems, Man, and Cybernetics, Tokyo, Japan* (Vol. 1, pp. 841–847), 12–15 Oct. 1999. Los Alamitos, CA: IEEE. https://doi.org/10.1109/icsmc.1999.814201.

Zuberek, W. M. (2009). Siphon-based verification of component compatibility. In W. Zamojski (Ed.), *2009 Fourth International Conference on Dependability of Computer Systems (DepCos-RELCOMEX'09), Brunów, Poland* (pp. 123–132), 30 June-2 July 2009. Piscataway, NJ: IEEE. https://doi.org/10.1109/depcos-relcomex.2009.35.

Chapter 8
Distributed Autonomous and Asynchronous Automata (DA^3)

In the practice of computer engineering, various forms of automata are used to express the behavior of concurrent components. There are two reasons: graphical representation and individual modeling of separate components. UML state diagrams are a good example (UML n.d.).

For graphical representation of distributed systems, and for simulation in terms of parallel system components, distributed automata DA^3 (D-triple A or DA-cubed) were invented. We claim that our distributed automata better describe parallelism and cooperation in real distributed environment (with full asynchrony) than those enumerated below.

Several different concepts are called in the literature "distributed automata".

- Automata on distributed alphabets, communicating on common letters, based on Zielonka's automata (Zielonka 1987). The automata are called *distributed automata* in many papers concerning the behavior of concurrent systems, some of them additionally equipped with real time clocks for temporal analysis with real-time constraints: (Krishnan 2000; Van Chieu and Van Hung 2010; Muscholl 2015; Diekert and Muscholl 2012; Mohalik and Ramanujam 2002). These automata are called *asynchronous* in Niebert (1995), Mukund (2012) and Sandholm and Schwartzbach (1997), although they perform actions (make transitions) asynchronously only if the input symbols are different. They perform synchronous moves on common input symbols (and it is the only common aspect of the automata). These automata should be called synchronous from our point of view. Alur's Timed Automata (Alur and Dill 1994) (sometimes called distributed (Krishnan 2000) are very close to Zielonka's automata, they are simply equipped with time constraints and time invariants. CSP processes are similar, synchronizing on ! and ? operations (sending and receiving, respectively) rather than on symbols of input alphabet. The advantage of CSP is to determine the direction of communication, which should be supplied informally in the case of Zielonka's automata.

© Springer Nature Switzerland AG 2020

W. B. Daszczuk, *Integrated Model of Distributed Systems*, Studies in Computational Intelligence 817, https://doi.org/10.1007/978-3-030-12835-7_8

- Close to Zielonka's automata are Büchi automata used for LTL model checking [for instance in Spin (Holzmann 1997)]. They differ in distinguishing some states as accepting. They are called *distributed automata* in Brim et al. (2006).
- Message Passing Automata (MPA, called *distributed automata* in Bollig and Leucker (2004, 2005)) are really distributed and asynchronous. They contain ordered sets on symbols waiting for acceptance, called buffers or queues.
- Pushdown *Distributed Automata* (PDA) are equipped with local stacks of input symbols (Balan 2009).
- The two previous cases (MPA and PDA) are combined in Enea et al. (2014; Madhusudan and Parlato 2011) and also called *distributed automata.*
- The automata that are synchronous in fact, with a central synchronizing server. In such an automaton, two independent actions are performed "simultaneously" by independent processors. Synchronization between the processors is explicitly performed by a centralized processor: the synchronizer (Huguet and Petit 1995; Petit 1993).
- Grammar systems—languages for description of parallel systems, generated by automata with certain interleaving rules. They are called *distributed automata* in (Krithivasan and Ramanujan 2013; Krithivasan et al. 1999; Fernau et al. 2001; Păun 1995).
- A single large automaton split into distributed parts called *distributed automata* (Gros-Desormeaux et al. 2008; Caillaud et al. 1997).

We introduce a new version of automata, equivalent to IMDS formalism. We call them *Distributed Autonomous, Asynchronous Automata*—DA^3 (*D-triple A* or *DA-cubed*) to distinguish them from all the listed formalisms, all called *distributed automata*. Our automata reflect the behavior of distributed components. The servers make decisions (execute actions) individually without any knowledge of other servers (autonomy) and messages are sent regardless of the states of target servers (asynchrony). As there are two views of the distributed system in IMDS, two forms of DA^3 were developed: Server-DA^3 and Agent-DA^3 (S-DA^3 and A-DA^3).

8.1 Server Automata S-DA^3

The IMDS system in the server view can be shown as a set of communicating automata S-DA^3 (Distributed Server Automata), similar to MPA (third in the above enumeration):

- Server states are *nodes* (we use *node* instead of *state* to avoid ambiguity) of corresponding automaton.
- The initial server state is the *initial node* of the automaton.
- Server process actions are *transitions* of the automaton.
- The automaton is Mealy-style (Dick and Yao 2014), labels of the transitions in the automaton are in a form extracted from the actions; an IMDS action $(m,p)\Lambda(m',p')$ is converted to the transition from p to p' with the label m/m'.

It is a triple (*node*, *transition label*, *node*): $(p, m/m', p')$; m is the *input symbol* conditioning the transition while m' is the *output symbol* produced on the transition; messages are $m = (a, s, r)$, $m' = (a, s', r')$.

- The automaton is equipped with an *input set*—a set of pending input symbols (messages), corresponding to a set of pending messages at the server. Firing the transition $(p, m/m', p')$ in the server s automaton retrieves the symbol m from the input set of this automaton and inserts the symbol m' to the input set of an automaton of the server s' appointed by m'. The *initial input set* consists of the initial messages of agents directed to this server.
- The special agent-terminating action $(m, p)\Lambda(p')$ is converted to the transition $(p, m/p')$ that does not produce an output symbol.

Formally, having the definition of S, A, V, R, P, P_{ini}, M, M_{ini} from IMDS (respectively: servers, agents, values, services, states, initial states, messages, initial messages), we have the set $\mathcal{Z}$ (reflected S) of n_S distributed sever automata $\mathcal{Z} = \{z_i, i = 1..n_S\}$, where n_S is the number of servers in the set S. The ith distributed server automaton is

$$z_i = (s_i, P_i, p_{0i}, F_i, X_i, X_{0i}), \text{ where:} \quad (8.1)$$

where:

- $s_i \in S$—ith server,
- $P_i = \{p_i \mid p_i = (s_i, v), v \in V_i\}$—the set of *nodes*; V_i is the set of values of the server s_i states,
- $p_{0i} = (s_i, v_{0si}) \in P_i \cap P_{ini}$—the *initial node* (the initial state of the server s_i),
- $F_i = \{(p_1, m/m', p_2), p_1, p_2 \in P_i, m = (a, s_i, r_i) \in M, m' = (a, s_j, r_j) \in M, a \in A, r_i, r_j \in R$ or $(p_1, m/m', p_2), p_1, p_2 \in P_i, m = (a, s_i, r_i) \in M, a \in A, r_i \in R\}$—the set of *transitions* (for an ordinary action and an agent-terminating action, respectively),
- $X_i \in 2\hat{}\{m = (a, s_i, r_i) \in M, a \in A, r_i \in R\}$—the *input set* (^ is used to denote powerset); X_i is the variable having a set value: the sender of the message (the transition delivering the message) inserts the element, the transition execution removes the element, according to the rules for $T_{\mathcal{Z}}$ below,
- $X_{0i} \in 2\hat{}\{m = (a, s_i, r_i) \in M_{ini}, a \in A, r_i \in R\}$—the *initial input set* (restricted to messages pending initially at the server s_i),
- $\forall_{m1,\ m2 \in Xi}\ m_1 = (a_1, s_i, r_1), m_2 = (a_2, s_i, r_2), m_1 \neq m_2 \Rightarrow a_1 \neq a_2$—the condition of IMDS that for any agent at most one message may exists in the configuration (achieved by construction).

In $\mathcal{Z}$, $\forall_{zi \in \mathcal{Z}} \forall_{(p1,\ m/m',\ p2) \in Fi}\ m' = (a, sj, rj), a \in A, s_j \in S, r_j \in R, z_j \in \mathcal{Z}$ (each output symbol is the input symbol of the automaton belonging to $\mathcal{Z}$).

The condition $\forall_{mi \in Xi,\ mj \in Xj}\ m_1 = (a_1, s_i, r_1), m_2 = (a_2, s_j, r_2), r_1, r_2 \in R, j \neq i \Rightarrow a_1 \neq a_2$ is obvious: for each agent at most one message can exists in the configuration (achieved by construction).

The semantics of $\mathcal{Z}$ is defined as *global node space* ($\{T_\mathcal{Z}\}$, $T_{\mathcal{Z}0}$, $nextT_\mathcal{Z}$), where $\{T_\mathcal{Z}\}$ is the set of *global nodes*, $T_{\mathcal{Z}0}$ is the *initial global node* and $nextT_\mathcal{Z}$ is the *transition relation*, defined as follows:

- The global node of $\mathcal{Z}$ is $T_\mathcal{Z} = ((p_1, X_1), (p_2, X_2), \ldots, (p_n, X_n))$ (current states and sets of pending messages of all servers).
- If in $T_\mathcal{Z}$ there exists a z_i in which (p_{i1}, X_i), $(p_{i1}, m/m', p_{i2}) \in F_i$, $m \in X_i$, $m' = (a, s_j, r_j)$ (message m' is directed to a server s_j) then the *possible next global node* $T_\mathcal{Z}'$ is:

$$T'_\mathcal{Z} = \left((p_k, X'_k) \;\middle|\; \begin{cases} X'_k = X_k \backslash \{m\} \cup \{m'\} & for\ i = k = j \\ X'_k = X_k \backslash \{m\} & for\ k = i \neq j \\ X'_k = X_k \cup \{m'\} & for\ k = j \neq i \\ X'_k = X_k & for\ i \neq k \neq j \end{cases}, k = 1 \ldots n\right) \tag{8.2}$$

(the automaton z_i changes its node to p_{i2}, all other automata preserve their nodes; the message m is extracted from the input set X_i of the automaton z_i, the next message m' is inserted into the input set X_j of the automaton z_j appointed by m', all other input sets remain unchanged; the special case is for ($i = k = j$, first case), where the server s_i sends a message to itself).
- If in $T_\mathcal{Z}$ there exists z_i in which (p_1, X_i), $(p_{i1}, m/, p_{i2}) \in F_i$, $m \in X_i$ (message m terminates the agent appointed by m) then the *possible next global node* is:

$$T'_\mathcal{Z} = \left((p_k, X'_k) \;\middle|\; \begin{cases} X'_k = X_k \backslash \{m\} & for\ k = i \\ X'_k = X_k & for\ k \neq i \end{cases}, k = 1 \ldots n\right) \tag{8.3}$$

(the automaton z_i changes its node to p_{i2}, all other automata preserve their nodes; the message m is extracted from the input set X_i of the automaton z_i, all other input sets remain unchanged).
- The initial global node is $T_{\mathcal{Z}0} = ((p_{01}, X_{01}), (p_{02}, X_{02}), \ldots, (p_{0n}, X_{0n}))$.
- For a given global node $T_\mathcal{Z}$, the transition relation $nextT_\mathcal{Z}(T_\mathcal{Z})$ is a set of pairs ($T_\mathcal{Z}$, $T_\mathcal{Z}'$). The transition relation $nextT_\mathcal{Z} = \bigcup_{T\mathcal{Z}} nextT_\mathcal{Z}(T_\mathcal{Z})$. If for $T_\mathcal{Z}$ there are many possible next global nodes, one of them is chosen in nondeterministic way.

The *global graph* of $\mathcal{Z}$ cooperation can be developed in such a way that the nodes are global nodes $T_\mathcal{Z}$, and the edges are transitions in automata z_i. Of course, this graph is analogous to the LTS of IMDS system. Global nodes contain current states of all servers, and pending messages of all not-terminated agents (but split to subsets pending at individual servers). The input symbol (message) of the transition belongs to the input set of one of servers in the source global node, while the output symbol (message) belongs the input set of one of servers in the target global node. A fragment of the global node space for the "strokes" system presented in Figs. 3.5 and 3.6 is presented in Fig. 8.1.

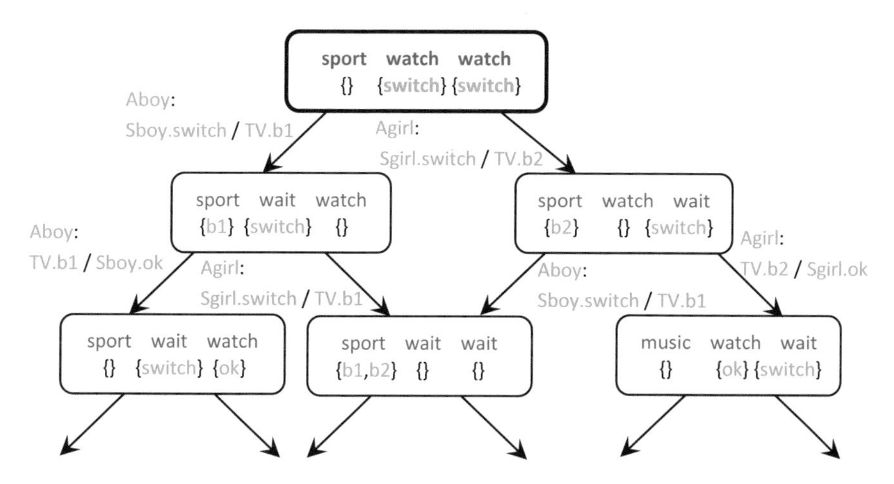

Fig. 8.1 A fragment of global node space of S-DA^3 automata for a "strokes". Global nodes are composed over `TV`, `Sgirl` and `Sboy` servers: first row contains states of the servers, second row input sets

It is simpler at the intuitive level, illustrated in Fig. 5.1 (for the "Two-semaphores" system—server view). Instead of individual server automata, the automata for server types are shown (with formal parameters of agents and servers used). Input sets of the automata $X_{proc[1]}$, $X_{proc[2]}$, $X_{sem[1]}$, $X_{sem[2]}$ *and* X_r are not shown in the picture, as they change during the operation of the automata. The agent formal parameters should be added for server types automata:

```
proc(A,sem[1..2]),
sem(A[1..2],proc[1..2]),
R(A3)
```

The S-DA^3 automata, being the instances of the server type automata, are equipped with actual parameters:

```
proc[1]:proc(A[1],sem[1..2]),
proc[2]:proc(A[2],sem[2],sem[1]),
sem[1]:sem(A[1..2],proc[1..2]),
sem[2]:sem(A[1..2],proc[1..2]),
r:r(A3).
```

Each automaton is equipped with the input set of pending messages, for example for the `sem[1]` automaton:

$X_{sem[1]}$ ∈ 2^{(`A[1]`, `sem[1]`, *up*),(`A[1]`, `sem[1]`, *down*),(`A[2]`, `sem[1]`, *up*),(`A[2]`, `sem[1]`, *down*)}

Note that the last condition in z_i requires that the former two messages cannot be included in $X_{sem[1]}$ at the same time, likewise the latter two messages.

The initial input sets are:

$X_{0\ sem[1]}=\varnothing$,
$X_{0\ sem[2]}=\varnothing$,
$X_{0\ proc[1]}=\{(\texttt{A[1],proc[1],start})\}$,
$X_{0\ proc[2]}=\{(\texttt{A[1],proc[2],start})\}$,
$X_{0\ r}=\{(\texttt{A3,r,res})\}$.

The S-DA^3 are similar to Message Passing Automata. The difference lies in ordering messages on the automaton input: in MPA pending messages are ordered in the input queue (or input buffer) (Bollig and Leucker 2004, 2005), while in S-DA^3 any message from the input set may cause a transition (no ordering). If input buffers are bounded, for example in implementation of distributed automata, a deadlock may occur due to all processes sending to full buffers. This is the case when the size of buffers is too small (Olson and Evans 2005). IMDS helps to overcome this problem by setting an exact limit on the maximum size of the input set (or input buffer): it is simply the number of agents.

8.2 Agent Automata A-DA^3

The IMDS system in the agent view can be shown as a set of communicating automata A-DA^3 (Agent Distributed Autonomous and Asynchronous Automata). We use the term *node* in these automata instead of *state*, because states are attributed to servers in IMDS and this may be misleading. The A-DA^3 automata are similar to Timed Automata with variables, as in Uppaal (Behrmann et al. 2006) (we do not use time dependencies of TA in the presentation of A-DA^3):

- Agent messages are the *nodes* of the corresponding automaton.
- The agent initial message of the is the *initial node* of the automaton.
- Agent process actions are the *transitions* of the automaton.
- The automaton is Mealy-style (Dick and Yao 2014); the labels of the transitions in the automaton have the form extracted from actions; the IMDS action $(m,p)\Lambda(m',p')$ is converted to a transition from m to m' with the label p/p'. It is a triple (*node, transition label, node*): $(m,p/p',m')$ [p is the *input symbol* conditioning the transition while p' is the *output symbol* produced on the transition; server states are $p = (s, v)$, $p' = (s, v')$].
- In the case of the agent-terminating action $(m,p)\Lambda(p')$, the special node $t_{\varepsilon i}$ in the automaton of ith agent is added as *terminating node*, and the transition is of the form $(m,p/p',t_{ei})$. For $t_{\varepsilon i}$ no outgoing transition is defined.
- The system is equipped with a *global input vector* (vector of global current input symbols), corresponding to the vector of current server states. Firing the transition $(m,p/p',m')$ in the automaton replaces the symbol p with the symbol p' in the vector. The *initial global input vector* consists of the initial states of all automata.

Formally, having the definition of S, A, V, R, P, P_{ini}, M, M_{ini} from IMDS (respectively: servers, agents, values, services, states, initial states, messages, initial messages), we have the set $\mho$ (reflected A, rounded to distinguish it from a universal quantifier) of n_A distributed agent automata $\mho = \{\mathfrak{v}_i,\ i = 1..n\}$, where n_A is the number of agents in the set A. The ith distributed agent automaton is:

$\mathfrak{v}_i = (a_i, M_i,\ m_{0i},\ t_i,\ G_i,\ Y,\ Y_0)$, where:

- a_i—ith agent,
- $M_i = \{m_i = (a_i, s_j, r_j),\ s_j \in S,\ r_j \in R_i\ or\ m_i = t_{\mathfrak{v}i}\}$—the set of *nodes* (messages, restricted to the agent a_i), $t_{\mathfrak{v}i}$ is a *terminating node* if the agent terminates (it appears on output of the terminating action),
- $m_{0i} = (a_i, s_j, r_{0j}) \in M_i \cap M_{ini}$—the *initial node* (the initial message of the agent a_i),
- $G_i = \{(m_1, p/p', m_2)$ or $(m_1, p/p', t_{\mathfrak{v}i}) \mid p = (s_1, v_1), p' = (s_2, v_2), m_1, m_2 \in M_i, s_1, s_2 \in S, v_1 \in V_{s1}, v_2 \in V_{s2}\}$—the set of *transitions*,
- $Y = [p_j = (s_j, v_{sj})],\ j = 1..n,\ s_j \in S,\ v_{sj} \in V_{sj}$—the *global input vector* (common for all $\mathfrak{v}_i$ in the system); Y is the vector of variables, every variable has the range over the set of values of the server s_j: the action changes the value of the variable at the position of its server, accordingly to the rules for the global node $T_{\mho}$ below; $Y[i]$ is the ith position of Y,
- $Y_0 = [p_{0j} = (s_j, v_{0sj})],\ j = 1...n,\ s_j \in S_{ini},\ v_{0sj} \in V_{sj}$—the *initial global input vector*, consisting of the initial states of all servers (common for all $\mathfrak{v}_i$ in the system).

The semantics of $\mho$ is defined as *global node space* ($\{T_{\mho}\}$, $T_{\mho 0}$, $nextT_{\mho}$), where $\{T_{\mho}\}$ is the set of *global nodes*, $T_{\mho 0}$ is the *initial global node* and $nextT_{\mho}$ is the *transition relation*, defined as follows:

- The global node of $\mho$ is $T_{\mho} = \{m, m', m'', \ldots, Y\}$ (of course, $\forall_{m,m' \in T_{\mho}}\ m = (a, s, r),\ m' = (a', s', r'),\ a \neq a'$)
- If in $T_{\mho}$ there exists $\mathfrak{v}_i$ in which m_i, $(m_i, p/p', m_j) \in G_i$, $m_i = (a_i, s_x, r_1)$, $m_j = (a_i, s_y, r_2) \in M$, $s_x, s_y \in S$, $r_1, r_2 \in R$, $p = Y[i]$, $p' = (s_x, v)$, $v \in V$ (message m_i causes the change of state p to p' in the server s_x appointed by m_i, and the message m_j to the server s_y is issued) then the *possible next global node* is:

$$T'_{\mho} : \forall_{m \in T_{\mho}} \begin{cases} m = m_i \wedge m = m_j \Rightarrow m_i \in T'_{\mho} \\ m = m_i \wedge m \neq m_j \Rightarrow m_j \in T'_{\mho}; \\ m \neq m_i \Rightarrow m_i \in T'_{\mho} \end{cases} \tag{8.4}$$
$$Y' = [p_1, \ldots, p_n] \mid p_k = \begin{cases} Y[k] & for\ k \neq x \\ p' & for\ k = x \end{cases}, k = 1 \ldots n$$

(automaton $\mathfrak{v}_i$ changes its node to m_j, all other automata preserve their nodes; state p in the input vector Y is replaced by p' in the position of the server s_x, appointed by p and p', all other elements of the vector Y remain unchanged).

- If in $T_{Ʉ}$ there exists $ɐ_i$ in which m_{i1}, $(m_i, p/p', t_{ɐi}) \in G_i$, $m_i = (a_i, s_x, r) \in M$, $s_x \in S$, $r \in R$, $p = Y[i]$, $p' = (s_x, v)$, $v \in V$ (message m_i is the last message in the run of the agent a_i, then the agent terminates, sever s_x appointed by the message m_i changes its state from p to p') then the *possible next global node* is:

$$T'_{Ʉ} : \forall_{m \in T_{Ʉ}} \begin{cases} m = m_i \Rightarrow m_i \notin T'_{Ʉ} \\ m \neq m_i \Rightarrow m_i \in T'_{Ʉ} \end{cases};$$
$$Y' = [p_1, \ldots, p_n] \mid p_k = \begin{cases} Y[k] & for\ k \neq x \\ p' & for\ k = x \end{cases}, k = 1 \ldots n] \tag{8.5}$$

(the automaton $ɐ_i$ changes its node to $t_{ɐi}$, all other automata preserve their nodes; the state p in Y is replaced by p' as above).
- The initial global node $T_{Ʉ0} = \{m_{01}, m_{02}, \ldots, m_{0n}, Y_0\}$.
- For a given global node $T_{Ƨ}$, the transition relation $nextT_{Ʉ}(T_{Ʉ})$ is a set of pairs $(T_{Ʉ}, T_{Ʉ}')$. The transition relation $nextT_{Ʉ} = \cup_{T_{Ʉ}} nextT_{Ʉ}(T_{Ʉ})$. In a given global node $T_{Ʉ}$, if there are multiple next nodes possible, one of them is chosen in nondeterministic way.

Distributed agent automata are illustrated in Figs. 8.2 and 8.3 (for the two-semaphore system—agent view). Instead of individual agent automata, the automata for agent types are shown (with formal parameters of used servers). For the instance of an automaton type, formal parameters should be replaced by actual instances of servers. Initial agents messages are in bold ovals. The terminating node ($t_{ɐ}$) is surrounded by a dashed oval. Agent identifies are omitted in message labels (nodes of the automata), because they are identical for all messages in given agent type automaton. For completeness, the global input vector of current server states should be added:

```
Y=[(sem[1],value∈{up,down}),
(sem[2],value∈{up,down}),
(proc[1],value∈{ini,first,sec,stop}),
(proc[2],value∈{ini,first,sec,stop}),
(r[1],value∈{res})].
```

The initial input vector is:

```
Y0=[(sem[1],up),(sem[2],up),(proc[1],ini),(proc[2],ini),
(r,res)].
```

Note that `A[1..2]` automata terminate, therefore "terminating node" t_A is used. There is no such node in `A3`, because this automaton is not expected to terminate.

The agent automata are instances of agent type automata with actual parameters:

```
A[1]:A(proc[1],sem[1,2]),
A[1]:A(proc[2],sem[2,1]),
A3:A3(r).
```

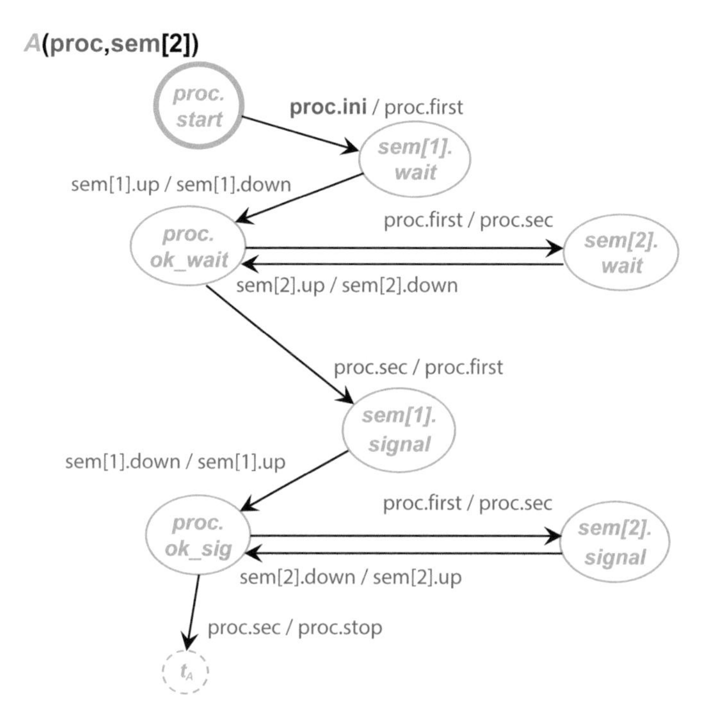

Fig. 8.2 Graphical representation of agent *A* automaton type

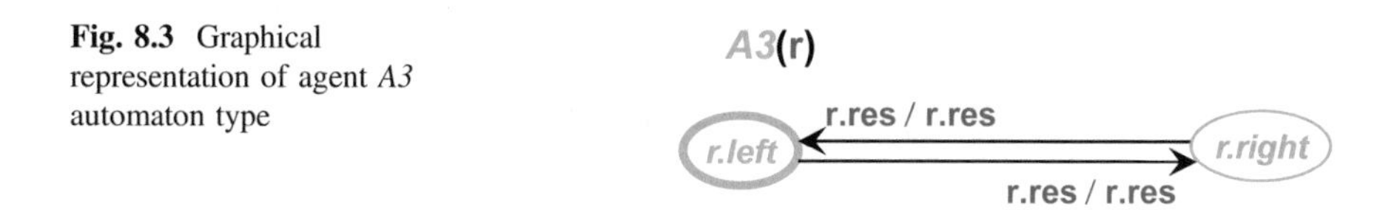

Fig. 8.3 Graphical representation of agent *A3* automaton type

A *global graph* of $\mathcal{U}$ cooperation can be developed analogously to the graph of $\mathcal{Z}$: the nodes of the global graph are global nodes $T_{\mathcal{U}}$, and the edges are transitions in the automata ε_i. This graph is analogous to the global graph of S-DA^3 and to the LTS of IMDS system (global nodes contain messages of all agents, input symbol/state of a transition should be attributed to the source global node, while output symbol-state to the target global node). The fragment of the global node space for the "strokes" system presented in Figs. 3.5 and 3.6 is presented in Fig. 8.4.

8.3 Using DA^3 in the Dedan Program

The basic form used in the Dedan program is the IMDS. However, the specification in the form of the relation between pairs$(state, message)\Lambda(state', message')$ is

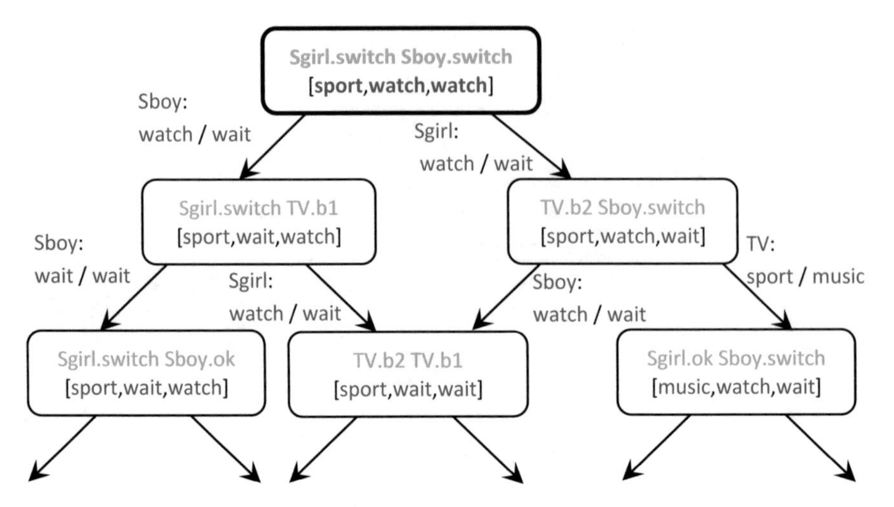

Fig. 8.4 A fragment of global node space of A-DA^3 automata for a "strokes" system in Figs. 3.2 and 3.3, in first row pending massages in a global node are shown, second row contains global input vector Y

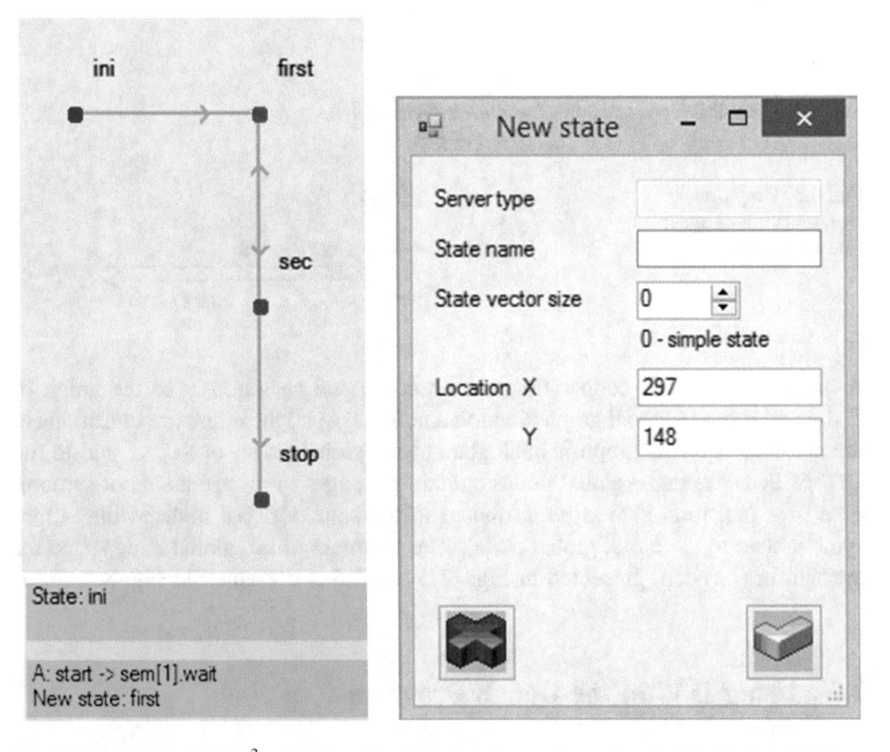

Fig. 8.5 A design of SDA^3 automaton. On the left, a shape of an automaton is displayed. Below, a chosen state and a list of transitions from this state are shown. On the right a window for parameters of new state are displayed

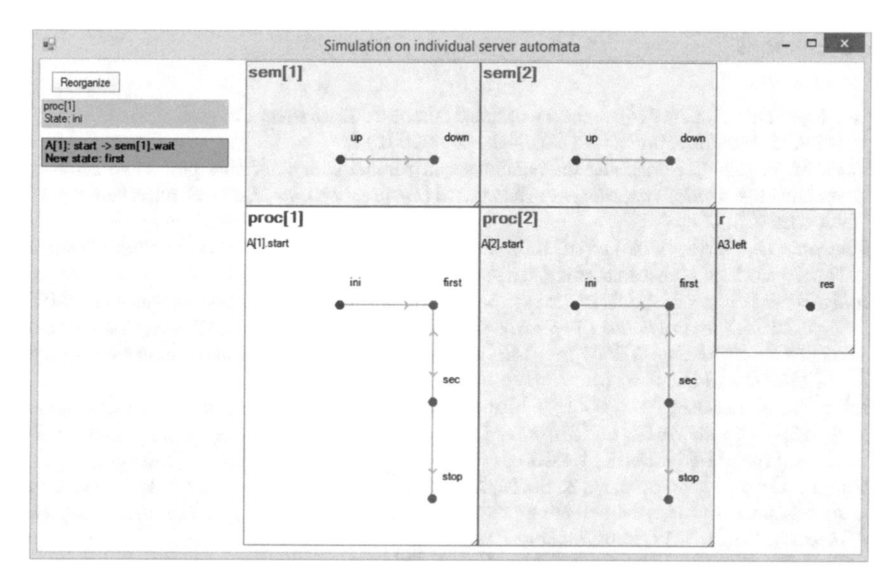

Fig. 8.6 Simulation over S-DA^3 automata. Every automaton has its own sub-window, a chosen automaton has blue caption and light background. On the left, current state of the chosen automaton and a list of possible transitions are displayed

exotic for users. Therefore, an alternative input form of DA^3 automata is provided. For example, Fig. 8.5 shows the graphic form of the automaton type (`proc` automaton type of the “2 semaphores” system) in the design window. The user can move the state positions, add states and transitions, delete them etc. In the figure, the window of a new state definition is shown on the right.

The system can be simulated over the global configuration space (the LTS), but it is also possible to simulate it in terms of S-DA^3, as illustrated in Fig. 8.6. All automata in the system are displayed, with input sets of pending messages shown under automata identifiers. The current states of the automata are blue.

The user can select an automaton (*proc[1]* in the example, the selected automaton has a white background and blue name), and on the left a list of transitions from the current state of the selected automaton is displayed (with prepared ones highlighted; it is only one transition in this case, and it is prepared). Then, the user can choose the transition from the prepared ones. In the example, this is only one transition prepared, leading from *ini* to *first* (with acceptance of `start` message and issuance of the *wait* message to *sem[1]*). If the user clicks the prepared transition, it is “executed” and the destination automaton of the message becomes current (*sem[1]* in this case).

References

Alur, R., & Dill, D. L. (1994). A theory of timed automata. *Theoretical Computer Science, 126*(2), 183–235. https://doi.org/10.1016/0304-3975(94)90010-8.

Balan, M. S. (2009). Serializing the parallelism in parallel communicating pushdown automata systems. *Electronic Proceedings in Theoretical Computer Science, 3,* 59–68. https://doi.org/10.4204/EPTCS.3.5.

Behrmann, G., David, A, & Larsen, K. G. (2006). *A tutorial on Uppaal 4.0*. Aalborg, Denmark. http://www.it.uu.se/research/group/darts/papers/texts/new-tutorial.pdf.

Bollig, B., & Leucker M. (2004). Message-passing automata are expressively equivalent to EMSO logic. In *15th International Conference CONCUR 2004—Concurrency Theory, London, UK, August 31–September 3, 2004* (pp. 146–160). Berlin, Heidelberg: Springer. https://doi.org/10.1007/978-3-540-28644-8_10.

Bollig, B., & Leucker, M. (2005). A hierarchy of implementable MSC languages. In *Formal techniques for networked and distributed systems—FORTE 2005, Taipei, Taiwan, October 2–5, 2005* (pp. 53–67). Berlin, Heidelberg: Springer. https://doi.org/10.1007/11562436_6.

Brim, L., Černá, I., Moravec, P., & Šimša, J. (2006). How to order vertices for distributed LTL model-checking based on accepting predecessors. *Electronic Notes in Theoretical Computer Science, 135*(2), 3–18. https://doi.org/10.1016/j.entcs.2005.10.015.

Caillaud, B., Caspi, P., Girault, A., & Jard, C. (1997). Distributing automata for asynchronous networks of processors. *European Journal of Automation (RAIRO-APII-JESA), 31*(3), 503–524. Url: ftp://ftp.inrialpes.fr/pub/bip/pub/girault/Publications/Jesa97/main.pdf

Dick, G., & Yao, X. (2014). Model representation and cooperative coevolution for finite-state machine evolution. In *2014 IEEE Congress on Evolutionary Computation (CEC), Beijing, China, July 6–11, 2014* (pp. 2700–2707). New York, NY: IEEE. https://doi.org/10.1109/cec.2014.6900622.

Diekert, V., & Muscholl, A. (2012). On distributed monitoring of asynchronous systems. In *19th International Workshop on Logic, Language, Information and Computation, WoLLIC 2012, Buenos Aires, Argentina, September 3–6, 2012* (pp. 70–84). Berlin, Heidelberg: Springer. https://doi.org/10.1007/978-3-642-32621-9_5.

Enea, C., Habermehl, P., & Parlato, G. (2014). On the path-width of integer linear programming. *Electronic Proceedings in Theoretical Computer Science, 161,* 74–87. https://doi.org/10.4204/EPTCS.161.9.

Fernau, H., Holzer, M., & Freund, R. (2001). Hybrid modes in cooperating distributed grammar systems: Internal versus external hybridization. *Theoretical Computer Science, 259*(1–2), 405–426. https://doi.org/10.1016/S0304-3975(00)00022-0.

Gros-Desormeaux, H., Fouchal, H., & Hunel, P. (2008). A comparison of distributed test generation techniques. In R. Ben Ayed & K. Djemame (Eds.), *Second International Conference on Verification and Evaluation of Computer and Communication Systems VECoS'08, Leeds, UK, July 2–3, 2008* (pp. 38–49). Swinton, UK: British Computer Society. https://dl.acm.org/citation.cfm?id=2227467.

Holzmann, G. J. (1997). The model checker SPIN. *IEEE Transactions on Software Engineering, 23*(5), 279–295. https://doi.org/10.1109/32.588521.

Huguet, S., & Petit, A. (1995). Modular constructions of distributing automata. In *Mathematical Foundations of Computer Science 1995, 20th International Symposium, MFCS'95, Prague, Czech Republic, August 28–September 1, 1995* (pp. 467–478). Berlin, Heidelberg: Springer. https://doi.org/10.1007/3-540-60246-1_152.

Krishnan, P. (2000). Distributed timed automata. *Electronic Notes in Theoretical Computer Science, 28,* 5–21. https://doi.org/10.1016/S1571-0661(05)80627-9.

Krithivasan, K., Balan, M. S., & Harsha, P. (1999). Distributed processing in automata. *International Journal of Foundations of Computer Science, 10*(04), 443–463. https://doi.org/10.1142/S0129054199000319.

Krithivasan, K., & Ramanujan, A. (2013). On the power of distributed bottom-up tree automata. *International Journal of Advanced Computer Science, 3*(4), 184–190. http://worldcomp-proceedings.com/proc/p2011/FCS2998.pdf.
Madhusudan, P., & Parlato, G. (2011). The tree width of auxiliary storage. In *Proceedings of the 38th Annual ACM SIGPLAN-SIGACT Symposium on Principles of Programming Languages—POPL'11, Austin, TX, January 26–28, 2011* (pp. 283–294). New York, NY, USA: ACM Press. https://doi.org/10.1145/1926385.1926419.
Mohalik, S., & Ramanujam, R. (2002). Distributed automata in an assumption-commitment framework. *Sadhana, 27*(2), 209–250. https://doi.org/10.1007/BF02717184.
Mukund, M. (2012). Automata on distributed alphabets. In *Modern applications of automata theory* (pp. 257–288). Co-Published with Indian Institute of Science (IISc), Bangalore, India. https://doi.org/10.1142/9789814271059_0009.
Muscholl, A. (2015). Automated synthesis of distributed controllers. In *Automata, Languages, and Programming—42nd International Colloquium, ICALP 2015, Kyoto, Japan, July 6–10, 2015, Part II* (pp. 11–27). https://doi.org/10.1007/978-3-662-47666-6_2.
Niebert, P. (1995). A ν-calculus with local views for systems of sequential agents. In *20th International Symposium on Mathematical Foundations of Computer Science MFCS'95, Prague, Czech Republic, August 28–September 1, 1995* (pp. 563–573). London, UK: Springer. https://doi.org/10.1007/3-540-60246-1_161.
Olson, A. G., Evans, B. L. (2005). Deadlock detection for distributed process networks. In *ICASSP'05. IEEE International Conference on Acoustics, Speech, and Signal Processing, Philadelphia, PA, March 18–23, 2005, Vol. V* (Vol. 5, pp. 73–76). New York, NY: IEEE. https://doi.org/10.1109/icassp.2005.1416243.
Păun G. (1995). Grammar systems: A grammatical approach to distribution and cooperation. In *Automata, languages and programming* (pp. 429–443). Newark, NJ: Gordon and Breach Science Publishers, Inc. https://doi.org/10.1007/3-540-60084-1_94.
Petit, A. (1993). Recognizable trace languages, distributed automata and the distribution problem. *Acta Informatica, 30*(1), 89–101. https://doi.org/10.1007/BF01200264.
Sandholm, A. B., & Schwartzbach, M. I. (1997). Distributed safety controllers for web services. *BRICS Report Series, 4*(47). https://doi.org/10.7146/brics.v4i47.19268.
UML. (n.d.). http://www.uml.org/
Van Chieu, D., & Van Hung, D. (2010). An extension of mazukiewicz traces and their applications in specification of real-time systems. In *2010 Second International Conference on Knowledge and Systems Engineering, Hanoi, Vietnam, October 7–9, 2010* (pp. 167–171). New York, NY: IEEE. https://doi.org/10.1109/kse.2010.39.
Zielonka, W. (1987). Notes on finite asynchronous automata. *RAIRO—Theoretical Informatics and Applications, 21*(2), 99–135. https://doi.org/10.1051/ita/1987210200991.

Chapter 9
Fairness in Distributed Systems Verification

Recall that in distributed systems three types of nondeterminism can be observed, modeled in IMDS:

- Nondeterminism between servers: modeling of distribution. Servers make their decisions independently, therefore if two or more servers have prepared actions (their states and pending messages match), each of them can execute its action first. This is the modeling of nondeterminism in real world of independently operating servers.
- Nondeterminism in server means that the current state of this server can match more than one pending message. The server should select the action to execute in nondeterministic way. Processes running in a server are scheduled somehow, it depends on an operating system policy (Pnueli and Sa'ar 2008) and run-time systems installed (like java (Laskowski et al. 2005) or.net (Zerzelidis and Wellings 2005) environments). Full nondeterminism is most general approximation of many possible scheduling policies.
- Nondeterminism in agent is a situation in which two or more actions of the agent are possible in the current state of the server appointed by the current message of the agent. This models the nondeterministic choice in computation. Loops are often modeled in such a way: if the loop ending condition is not modeled, the decision to continue the loop or to escape from the loop is taken in nondeterministic way. Nondeterminism in a single agent seldom model actual nondeterministic choices in a program. Usually, nondeterminism models abstraction from detailed decision making. However, there may be real nondeterministic choices, for example modes of operation (as in the bounded buffer system in Fig. 7.8), of activities selected by human users.

In all three cases, the nondeterministic choice should be fair, which means that if the same nondeterministic choice is to be decided infinite number of times, every option should be selected at least once. The verifier of the system should be fair in any one of the three types of nondeterministic choices: between servers, in server and in agent.

© Springer Nature Switzerland AG 2020

W. B. Daszczuk, *Integrated Model of Distributed Systems*, Studies in Computational Intelligence 817, https://doi.org/10.1007/978-3-030-12835-7_9

In model checking, several notions of fairness are introduced, mainly to model independence of parallel processes and fair scheduling in operating systems. The basic notions are justice (weak fairness) and compassion (strong fairness) (Kesten et al. 1998; Pnueli and Sa'ar 2008; Baier and Katoen 2008; Rozier 2011). A just scheduler executes the process infinitely often, while a compassionate scheduler ensures that a process which is enabled infinitely often is executed infinitely often.

Justice and compassion can be attributes of the entire verified system. Alternatively, to avoid computations in which a given transition is forever ignored, transitions can be marked as just or compassionate (Gómez and Bowman 2005). Compassion applied to a given pair of events p, q can be formulated as a temporal formula, which says that whenever event p occurs, subsequent event q must occur in the future:

- LTL : $\Box(p \Rightarrow \Diamond q)$ (9.1)
- CTL : $\mathbf{AG}(p \Rightarrow \mathbf{AF}\, q)$

9.1 The Benchmark—Bounded Buffer

The Dedan program was designed for specification and verification of distributed systems. Temporal verification for deadlock detection was planned using external model checkers: Spin (Holzmann 1997) for LTL (linear time logic (Clarke et al. 1999)), NuSMV (Cimatti et al. 2000) for CTL (computation tree logic (Clarke et al. 1999)) and Uppaal (Behrmann, David, et al. 2006) for TCTL (CTL and timed CTL, TCTL). The choice of the verifiers was based on the two aspects: they are popular and free (at least for academic use). Dedan converts the IMDS specification to input formats of the three programs: Promela, SMV and Timed Automata XML. The counterexamples generated in the case of a deadlock found are read to the Dedan program and presented in a readable form similar to sequence diagrams.

In the IMDS, several tests were developed to check the operation of the Dedan program, including the presented two-semaphore model, intersection, several versions of bounded buffer, dining philosophers, etc. In most examples everything worked, but one example caused false positive in deadlock detection. It was a version of bounded buffer.

The basic example of a bounded buffer system consists of K-element buffer and two sets of users: producers and consumers. The correct specification does not lead to deadlocks. Among several versions of bounded buffer, in one of them users were modified to play roles of both producers and consumers. Such a system is illustrated as a Petri net in Fig. 7.8. The action of production/consumption is selected in nondeterministic way (lines 14 and 15 in the source code below). In such a system, a deadlock may occur if all users decide to read from an empty buffer, or if all users decide to write to a full buffer. There is no user to put an element into a buffer in the former case, and to get an element and make a room in a buffer in the latter case.

To avoid such situations, the concept of butlers from Dijkstra's solution of dining philosophers problem (Scholten and Dijkstra 1982) was used (line 25). But there is a difference: in the Dijkstra's solution the butlers prevent inactively all the philosophers from taking their left forks and from taking their right forks (the last philosopher simply waits). In our solution, the butlers cannot hold the users because the latter should make the other decision: to produce instead of consuming or reverse (l.16,17).

Such a system should not fall into a deadlock if it is compassionate: after many attempts to select putting and being rejected (l.14,16), the user must at last choose to get (l.15) (and vice versa). Unfortunately, all three external verifiers failed to give correct result—they report a deadlock in the sense of the IMDS deadlock definition in the server view. This means that the verifiers report some states in the system in which messages are pending at the servers but they will never be served. In the following sections we show this vulnerability in all the three external verifiers.

The model in the IMDS notation is given below. For the compactness, identifiers are shortened: servers *Sprodcons* to *S*, *get_butler* to *get_b*, *put_butler* to *put_b*, agent *Aprodcons* to *A*. For readability, the vector *elem* is indexed from 0 (in fact, in Dedan indexing starts with 1 and additional state *elem_0* should be added).

```
1. #DEFINE N 3
2. #DEFINE K 3

3. server:   buf (agents A[N]; servers S[N]),
4. services  {put, get},
5. states    {elem[0..K]},
6. actions   {
7. <i=1..N>  <j=0..K-1>  {A[i].buf.put, buf.elem[j]}   ->
                          {A[i].S[i].ok_put, buf.elem[j+1]},
8. <i=1..N>  <j=1..K>    {A[i].buf.get, buf.elem[j]}   ->
                          {A[i].S[i].ok_get, buf.elem[j-1]},
9. };

10. server:   S (agents A;
                 servers buf,put_b:butler,get_b:butler),
11. services  {doSth,may_do,cannot_do,ok_put,ok_get,go_neutral},
12. states    {neutral,prod,cons},
13. actions   {
14.           {A.S.doSth, S.neutral}  -> {A.put_b.wants_do, S.prod}
15.           {A.S.doSth, S.neutral}  -> {A.get_b.wants_do, S.cons}
16.           {A.S.cannot_do, S.prod} -> {A.S.doSth, S.neutral}
17.           {A.S.cannot_do, S.cons} -> {A.S.doSth, S.neutral}
18.           {A.S.may_do, S.prod}    -> {A.buf.put, S.prod}
19.           {A.S.may_do, S.cons}    -> {A.buf.get, S.cons}
20.           {A.S.ok_put, S.prod}    -> {A.put_b.done, S.prod}
21.           {A.S.ok_get, S.cons}       -> {A.get_b.done, S.cons}
22.           {A.S.go_neutral, S.prod}  -> {A.S.doSth, S.neutral}
23.           {A.S.go_neutral, S.cons}  -> {A.S.doSth, S.neutral}
24. };

25. server:   butler (agents A[N]; servers S[N]),
26. services  {wants_do,done}
27. states    {
28. <i=1..N>  <j=0..N-2> {A[i].butler.wants_do, butler.want[j]} ->
                            {A[i].S[i].may_do, butler.want[j+1]}
29. <i=1..N>             {A[i].butler.wants_do, butler.want[N-1]} ->
                          {A[i].S[i].cannot_do, butler.want[N-1]}
30. <i=1..N>  <j=1..N-1> {A[i].butler.done, butler.want[j]} ->
                           {A[i].S[i].go_neutral, butler.want[j-1]}
31. };
32. servers   buf,S[N],get_b:butler,put_b:butler;
```

```
33. agents     A[N];

34. init        ->        {
35. <j=1..N>  S[j](A[j],buf,put_b,get_b).neutral,
36.           buf(A[1..N],S[1..N]).elem[0],
37.           get_b(A[1..N],S[1..N]).want[0],
38.           put_b(A[1..N],S[1..N]).want[0],

39. <j=1..N>  A[j].S[j].doSth,
40. }.
```

For illustration of the model, S-DA[3] automata are presented in Fig. 9.1. The names of the automata implementing server types are inside rounded boxes. Initial states of the automata are surrounded by bold ellipses. Some transitions are with repeaters of the form `<i=1..N>`. Multiple transitions (but not those with repeaters) have double arrows and double labels. Transitions implement the IMDS actions in such a way that input and output states of the transition are automaton nodes, the input message is the input symbol enabling the transition while the output message (if present) is the output symbol of the transition (automata are Mealy-style (Dick and Yao 2014)). Output messages are directed to the given servers, but they may not be accepted immediately —the messages can be pending in IMDS (due to interleaving, if some other server executes an action or simply the message does not match the current server state).

All are three types of nondeterminism in IMDS take place in the described model:

- Nondeterminism between servers takes place in any system consisting of more than one server, and we have many servers.
- Nondeterminism in server occurs, for example in the butler: it can have more than one request (`wants_do` message) and should grant the access to the buffer fairly.
- Nondeterminism in agent is modeled in the decision of each agent working on its own server: to produce or to consume.

For the purpose of model checking with external verifiers, the temporal formulas are slightly modified. In original version, E_s and E_a atomic formulas were specified as "an action is prepared (a state and a message match)". It is enough for the internal model checker, as it is designed for supporting fairness and the action which is prepared many times, must be fired. To be independent in the way of evaluating temporal formulas, E_s means "action is executed in server s" instead of "at least one action is prepared in the server s". Likewise, E_a means "action is executed with the agent's a message on input" instead of "action is prepared with the agent's a message".

For temporal verification of the example system, strong fairness (compassion) is needed. If the agent periodically tries to perform `get`, and it is consequently rejected, it should finally switch to `put` (because both `get` and `put` are possible from the `neutral` state of its home server).

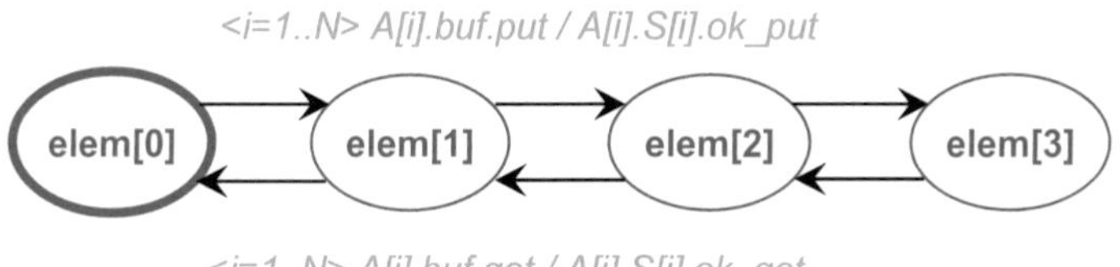

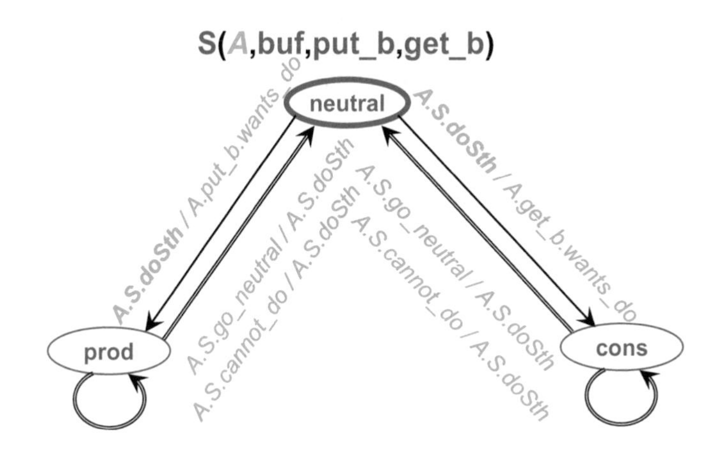

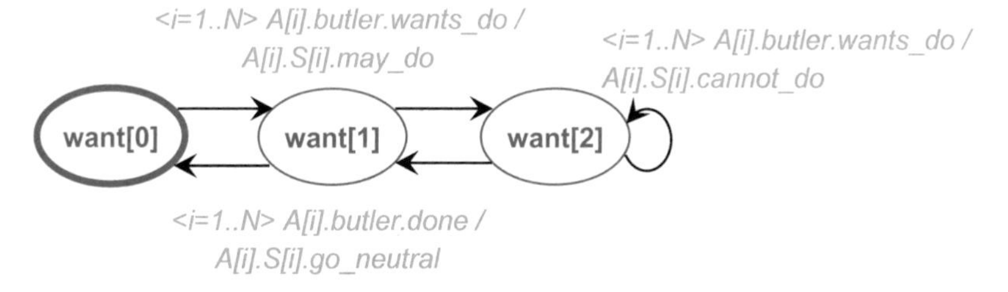

Fig. 9.1 Automata model of IMDS bounded buffer system

9.2 LTL Model Checking in Spin

For verification in linear temporal logic (LTL), the commonly used model checker Spin (Holzmann 1997) was used. Deadlock detection formulas are:

- communication deadlock in server s: $\Diamond\Box\ (D_s \wedge \neg\ E_s)$; the formula reads: "eventually always there will be messages pending at the server s, but no action will be executed from this time on",

- resource deadlock in agent a: $\Diamond\Box\ (D_a \wedge \neg E_a)$; "eventually always the pending message of the agent a will not cause any action to be executed",
- termination of agent a: $\Diamond\ (F_a)$; "eventually the agent a terminates".

The model is converted from the IMDS notation to Promela (Holzmann 1995), the input form of Spin. Below is a fragment of the Promela code for the server `buf` (the version of capacity 2, line 10). In the `do` statement (l.16) all actions of the server `buf` are implemented. Pending messages are stored in variables `buf_Aprodcons_1_service` and `buf_Aprodcons_2_service` (l.17,18). Execution of the action is signaled by setting the variable `buf_action` to `true`, then to `false` (for example in line 19). The definitions of servers `butler` and `S` are similar. After the definition of `buf`, the LTL formula for deadlock detection in the `buf` server is given (l.29).

```
1. //enumeration type of states and services of all servers:
2. mtype {none, put, get, elem_0, elem_1, elem_2, doSth, may_do, cannot_do, ok_put,
           ok_get, go_neutral, neutral, prod, cons, wants_do, done, want_0, want_1};
3. //buf_services = {put,get};
4. //buf_states = {elem_0,elem_1,elem_2};

5. mtype buf_Aprodcons_1_service,buf_Aprodcons_2_service;
6. bool buf_action;
7. mtype buf_state;
8. #define buf_c_Aprodcons_1 (buf_Aprodcons_1_service!=none &&
           buf_Aprodcons_1_service!=0)
9. #define buf_c_Aprodcons_2 (buf_Aprodcons_2_service!=none &&
           buf_Aprodcons_2_service!=0)

10. proctype buf(byte num; byte inistate; chan Aprodcons_1_Sprodcons_1,
        Aprodcons_2_Sprodcons_2, Aprodcons_1, Aprodcons_2)
11. {     mtype mes;
12.       buf_state=inistate;
13.       buf_Aprodcons_1_service=none;
14.       buf_Aprodcons_2_service=none;
15.       buf_action=0;
16.       do
17.           :: Aprodcons_1?mes -> buf_Aprodcons_1_service=mes;
18.           :: Aprodcons_2?mes -> buf_Aprodcons_2_service=mes;
19.           :: (buf_Aprodcons_1_service==put) && (buf_state==elem_0) -> buf_action=1;
                 buf_state=elem_1; buf_Aprodcons_1_service=none; buf_action=0;
                 Aprodcons_1_Sprodcons_1!ok_put
20.           :: (buf_Aprodcons_1_service==put) && (buf_state==elem_1) -> buf_action=1;
                 buf_state=elem_2; buf_Aprodcons_1_service=none; buf_action=0;
                 Aprodcons_1_Sprodcons_1!ok_put
21.           :: (buf_Aprodcons_1_service==get) && (buf_state==elem_2) -> buf_action=1;
                 buf_state=elem_1; buf_Aprodcons_1_service=none; buf_action=0;
                 Aprodcons_1_Sprodcons_1!ok_get
22.           :: (buf_Aprodcons_1_service==get) && (buf_state==elem_1) -> buf_action=1;
                 buf_state=elem_0; buf_Aprodcons_1_service=none; buf_action=0;
                 Aprodcons_1_Sprodcons_1!ok_get
23.           :: (buf_Aprodcons_2_service==put) && (buf_state==elem_0) -> buf_action=1;
                 buf_state=elem_1; buf_Aprodcons_2_service=none; buf_action=0;
                 Aprodcons_2_Sprodcons_2!ok_put
24.           :: (buf_Aprodcons_2_service==put) && (buf_state==elem_1) -> buf_action=1;
                 buf_state=elem_2; buf_Aprodcons_2_service=none; buf_action=0;
                 Aprodcons_2_Sprodcons_2!ok_put
25.           :: (buf_Aprodcons_2_service==get) && (buf_state==elem_2) -> buf_action=1;
                 buf_state=elem_1; buf_Aprodcons_2_service=none; buf_action=0;
                 Aprodcons_2_Sprodcons_2!ok_get
26.           :: (buf_Aprodcons_2_service==get) && (buf_state==elem_1) -> buf_action=1;
                 buf_state=elem_0; buf_Aprodcons_2_service=none; buf_action=0;
                 Aprodcons_2_Sprodcons_2!ok_get
27.       od
28. }

29. ltl buf_dd {!(<>[]((buf_c_Aprodcons_1 || buf_c_Aprodcons_2) && buf_action==0))}
```

At the end of the code, server variables are declared and initialized (line 7 in the code below for the server `buf`). For each pair (*agent, target server of its messages*) the communication channel of capacity 1 is declared (channel `A_1_buf` for messages sent by the agent `A_1` to the server `buf`, line 2). These channels are actual parameters of server variables (Promela processes—lines 10-14). The initial states are also sent as parameters to the server processes, and the initial agent messages are placed in the appropriate channels in lines 15 and 16 (all other channels are initially empty).

```
1. init   {
2.        chan Aprodcons_1_buf = [1] of {mtype};
3.        chan Aprodcons_2_buf = [1] of {mtype};
4.        chan Aprodcons_1_Sprodcons_1 = [1] of {mtype};
5.        chan Aprodcons_2_Sprodcons_2 = [1] of {mtype};
6.        chan Aprodcons_1_get_butler = [1] of {mtype};
7.        chan Aprodcons_2_get_butler = [1] of {mtype};
8.        chan Aprodcons_1_put_butler = [1] of {mtype};
9.        chan Aprodcons_2_put_butler = [1] of {mtype};

10. run buf(1, elem_0, Aprodcons_1_Sprodcons_1, Aprodcons_2_Sprodcons_2,
          Aprodcons_1_buf, Aprodcons_2_buf);
11. run Sprodcons(1, neutral,Aprodcons_1_put_butler, Aprodcons_1_get_butler,
          Aprodcons_1_buf, Aprodcons_1_Sprodcons_1);
12. run Sprodcons(2, neutral, Aprodcons_2_put_butler, Aprodcons_2_get_butler,
          Aprodcons_2_buf, Aprodcons_2_Sprodcons_2);
13. run butler(1, want_0, Aprodcons_1_Sprodcons_1, Aprodcons_2_Sprodcons_2,
          Aprodcons_1_get_butler, Aprodcons_2_get_butler);
14. run butler(2, want_0, Aprodcons_1_Sprodcons_1, Aprodcons_2_Sprodcons_2,
          Aprodcons_1_put_butler, Aprodcons_2_put_butler);
15. Aprodcons_1_Sprodcons_1!doSth;
16. Aprodcons_2_Sprodcons_2!doSth;
17. }
```

Spin gives the answer *true* to the question about deadlock. Spin has two versions of dealing with fairness: there are no fairness and weak fairness (they affect the time of verification). In both versions a deadlock is reported. Its graphical representation in Dedan sequence diagram is presented in Fig. 9.2. The *false positive* is identified when one user sent *put* (the putting action is not completed yet) and the other endlessly asks the `put_butler` to grant for putting. The butler endlessly responds `cannot_do`, and the user never switches to getting. Two kinds of fairness are broken: between servers (if the putting action is finished, it would allow the other user to put) and in agent (infinitely choosing production, never switching to consumption).

9.3 CTL and LTL Model Checking in NuSMV

The same system was modeled in the language of SMV, the input form of the NuSMV verifier (Cimatti et al. 2000), for CTL verification. The CTL versions of deadlock and termination detection formulas are as follows:

- communication deadlock in the server *s:* **EF AG** $(D_s \wedge \neg E_s)$
- resource deadlock in the agent *a:* **EF AG** $(D_a \wedge \neg E_a)$
- termination of the agent *a:* **AF** (F_a)

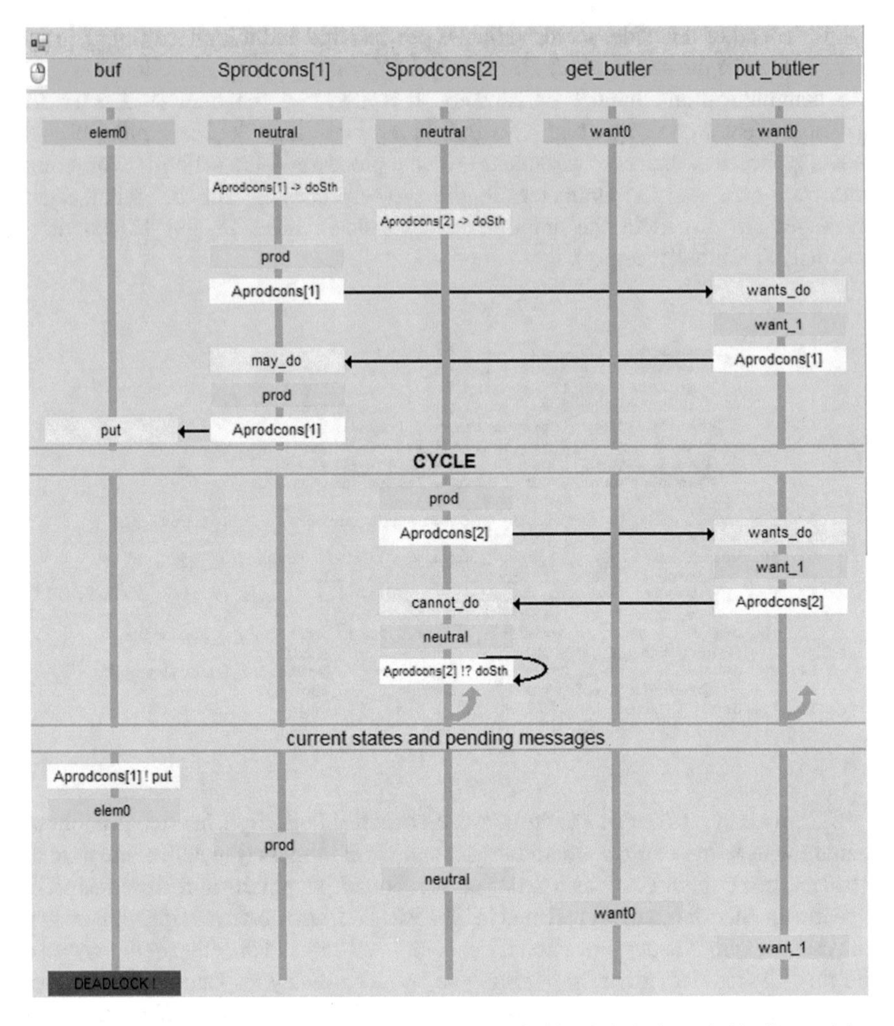

Fig. 9.2 Sequence diagram of the false deadlock identified by Spin in bounded buffer model

The meaning of the formulas is similar to LTL.

Modeling in SMV is based on changing value of the variables. On the example of the `buf` server in the two-user, two-element version of the system in SMV:

- the variable `state:{elem_0,elem_1,elem_2}` (line 7) storing the server state (initialized to the initial value `inistate`),
- for each agent—the pending message: the variables `A_1_service:{none, put,get}` and `A_2_service` (lines 3,4),
- the variable `phase:{rec,A_1_send,A_2_send,change}` (line 8) defines the action execution phase (receiving messages and selecting the agent

to perform the action—*rec*, selecting the action between prepared actions of the selected agent—*A_1_send, A_2_send*, sending the output message and changing the server state—*change*),

- for every agent – the message to be sent (*A_1_to_S_1_mes, A_2_to_S_2_mes* – lines 5 and 6).

The next lines contain actions that change the values of the variables according to current values.

```
1. MODULE buf (inistate,A_1_mes,A_2_mes)

2. VAR
3.          A_1_service: {none,put,get};
4.          A_2_service: {none,put,get};
5.          A_1_to_Sprodcons_1_mes: {none,doSth,ok_put,ok_get};
6.          A_2_to_Sprodcons_2_mes: {none,doSth,ok_put,ok_get};
7.          state: {elem_0,elem_1,elem_2};
8.          phase: {rec,A_1_send,A_2_send,change};

9. ASSIGN
10.         init(A_1_service):=none;
11.         next(A_1_service):=case
12.             phase=rec & A_1_mes=put: {none,put};
13.             phase=rec & A_1_mes=get: {none,get};
14.             phase=change & A_1_to_Sprodcons_1_mes!=none: none;
15.             TRUE: A_1_service;
16.         esac;
17.         init(A_2_service):=none;
18.         next(A_2_service):=case
19.             phase=rec & A_2_mes=put: {none,put};
20.             phase=rec & A_2_mes=get: {none,get};
21.             phase=change & A_2_to_Sprodcons_2_mes!=none: none;
22.             TRUE: A_2_service;
23.         esac;
24.         init(phase):=rec;
25.         next(phase):=case
26.           phase=rec & A_1_service=put & state=elem_0: A_1_send;
27.           phase=rec & A_1_service=put & state=elem_1: A_1_send;
28.           phase=rec & A_1_service=get & state=elem_2: A_1_send;
29.           phase=rec & A_1_service=get & state=elem_1: A_1_send;
30.           phase=rec & A_2_service=put & state=elem_0: A_2_send;
31.           phase=rec & A_2_service=put & state=elem_1: A_2_send;
32.           phase=rec & A_2_service=get & state=elem_2: A_2_send;
33.           phase=rec & A_2_service=get & state=elem_1: A_2_send;
34.           phase=rec & A_1_service=none & A_2_service=none: rec;
```

```
35.          phase=A_1_send & A_1_to_Sprodcons_1_mes!=none: change;
36.          phase=A_2_send & A_2_to_Sprodcons_2_mes!=none: change;
37.          phase=change: rec;
38.          TRUE: phase;
39.        esac;
40.        init(A_1_to_Sprodcons_1_mes):=none;
41.        next(A_1_to_Sprodcons_1_mes):=case
42.         phase=A_1_send & A_1_service=put & state=elem_0: ok_put;
43.         phase=A_1_send & A_1_service=put & state=elem_1: ok_put;
44.         phase=A_1_send & A_1_service=get & state=elem_2: ok_get;
45.         phase=A_1_send & A_1_service=get & state=elem_1: ok_get;
46.         TRUE: none;
47.      esac;
48.      init(A_2_to_Sprodcons_2_mes):=none;
49.      next(A_2_to_Sprodcons_2_mes):=case
50.         phase=A_2_send & A_2_service=put & state=elem_0: ok_put;
51.         phase=A_2_send & A_2_service=put & state=elem_1: ok_put;
52.         phase=A_2_send & A_2_service=get & state=elem_2: ok_get;
53.         phase=A_2_send & A_2_service=get & state=elem_1: ok_get;
54.         TRUE: none;
55.      esac;
56.      init(state):=inistate;
57.      next(state):=case
58.         phase=change & A_1_service=put & state=elem_0: elem_1;
59.         phase=change & A_1_service=put & state=elem_1: elem_2;
60.         phase=change & A_1_service=get & state=elem_2: elem_1;
61.         phase=change & A_1_service=get & state=elem_1: elem_0;
62.         phase=change & A_2_service=put & state=elem_0: elem_1;
63.         phase=change & A_2_service=put & state=elem_1: elem_2;
64.         phase=change & A_2_service=get & state=elem_2: elem_1;
65.         phase=change & A_2_service=get & state=elem_1: elem_0;
66.         TRUE: state;
67.      esac;
```

The formal parameters of the server (line 1) are: initial state and channels of two agents sending messages to *buf*.

Communication channels are not included in SMV language and should be modeled, for example the *buf* input channel is the following (the variable *state* is the message directed to the server *buf*, *input* is used to send a message, *output* is for receiving and cleaning the channel):

```
1. MODULE chan_buf (inimes,input,output)
2. VAR
3.          state: {none,put,get};
```

```
 4. ASSIGN
 5.         init(state):=inimes;
 6.         next(state) := case
 7.              input!=none & state=none: input;
 8.              output!=none & state!=none: none;
 9.              TRUE: state;
10.         esac;
```

Declaration of variables and their initialization are in the *main* module, channels first (the former two with initial messages - lines 3,4), and then servers (lines 7–9):

```
1. MODULE main
2. VAR
3.          A_1_proc_1 : chan_proc(doSth, proc_1.A_to_proc_mes,
                 buf.A_1_to_Sprodcons_1_mes, proc_1.A_service);
4.          A_2_proc_2 : chan_proc(doSth, proc_2.A_to_proc_mes,
                 buf.A_2_to_Sprodcons_2_mes, proc_2.A_service);

5.          A_1_buf : chan_buf(none, proc_1.A_to_buf_mes, buf.
                 A_1_service);
6.          A_2_buf : chan_buf(none, proc_2.A_to_buf_mes, buf.
                 A_2_service);

7.          proc_1 : proc(neutral, A_1_proc_1.state);
8.          proc_2 : proc(neutral, A_2_proc_2.state);
9.          buf : buf(elem_0, A_1_buf.state, A_2_buf.state);
```

The example of the NuSMV CTL formula for deadlock detection—equivalent to $\mathbf{EF}\ \mathbf{AG}(D_s \wedge \neg E_s)$—is for the example of server *buf* (if the message is received from *A_1* or *A_2*, it will be proceeded when the server leaves the *rec* phase):

```
NAME buf_dd := AG ((buf.A_1_service!=none | buf.A_2_service!=none)
         -> AF buf.phase!=rec)
```

Fairness between servers is natural, but fairness in server and in agent should be modeled explicitly by *COMPASSION* commands. This command specifies which options should be selected at least once if enabled infinitely frequently. The example shows the statements for nondeterminism in sever, when the message *put* from *A_1* is pending in *rec* phase infinitely often, and then it must eventually cause the next message to be sent in the context of *A_1* (the phase changes to *A_1_send*):

```
COMPASSION
        (phase=rec & A_1_mes=put & state=elem_0, phase = A_1_send)
COMPASSION
        (phase=rec & A_2_mes=put & state=elem_0, phase = A_2_send)
```

This looks appealing, but after invocation of NuSMV, the message says that *COMPASSION* statements do not guarantee to be held in CTL verification. Indeed, the result is similar to that of Spin:

```
WARNING *** The model contains COMPASSION declarations.         ***
WARNING *** Full fairness is not yet fully supported in NuSMV.  ***
WARNING *** Currently, COMPASSION declarations are only         ***
WARNING *** supported for BDD-based LTL Model Checking.         ***
WARNING *** Results of CTL Model Checking and of Bounded        ***
WARNING *** Model Checking may be wrong.                        ***
```

Following the advice above, we tried to verify in LTL, which is also possible in NuSMV. The example LTL formula is:

```
LTLSPEC    NAME A_1_dd := G ((buf.A_1_service!=none | proc_1.A_service!=none)
        -> F (buf.A_1_service=none & proc_1.A_service=none))
```

The result is even worse, because in LTL the verification seems to last infinitely and we interrupted it. The probable reason is that compassion requirements are usually included in the formulas, on the basis of: $(\varphi_1 \wedge \varphi_2 \wedge \ldots) \Rightarrow \psi$ ((Katoen 1999), Sect. 3.9), where φ_1, φ_2, … are compassion requirements supplied by the designer in *COMPASSION* statements, and ψ is the target formula specified in *LTLSPEC* statement. There are 21 *COMPASSION* statements in the specification, which results in very long, unacceptable verification time.

9.4 CTL Model Checking in Uppaal

The last external model checker used was Uppaal (Behrmann, David, et al. 2006). The verifier is designed to check CTL formulas with real-time constraints (TCTL). In this analysis, we do not use time constraints, therefore the model and verification concern “ordinary” CTL (without time constraints).

The code for `buf` server in the input form of Uppaal (XML Timed Automata) is presented below. The model is in the form of an automaton, with Uppaal locations being the `buf` states, and Uppaal transitions being the actions. For each server, the enumeration type of its services is defined (for example type `buf_service` for the server `buf`—line 6 and next lines with individual variables). The channels between the servers are implemented as simple variables of the respective types, holding a single message or being empty—the value *none* (line 2). The automaton is given the channels for message exchange with other servers as the parameters (separate for each agent—l.14). The additional parameter *myNum* (l.14) is the number of the automaton instance within the automaton template (server type). Server states are implemented as locations (l.17-19). Every action has the form a *transition*, connecting the locations with the *guard* label being a message pending at the channel output to the target server (l.24). The channels are implemented as simple variables holding single message. Execution of the transition causes the "consumption" of the input message (l.26) and issuing the output message (l.27).

```
 1. <declaration>
 2. const int none=0;

 3. typedef int[0,2] buf_service;
 4. const int buf_put=1;
 5. const int buf_get=2;

 6. typedef int[0,6] S_service;
 7. const int S_doSth=1;
 8. //next services of S are declared here: S_may_do=2, S_cannot_do=3, ...
 9. //then type butler_service with two services
10. </declaration>

11. <template>
12. <name>buf</name>
13. <parameter>
14. int myNum, Sprodcons_service &buf_Aprodcons_1_Sprodcons_1,
      Sprodcons_service &buf_Aprodcons_2_Sprodcons_2, buf_service
      &Aprodcons_1_buf, buf_service &Aprodcons_2_buf
15. </parameter>
16. <declaration />
17. <location id="id0"> <name>elem_0</name> </location>
18. <location id="id1"> <name>elem_1</name> </location>
19. <location id="id2"> <name>elem_2</name> </location>
20. <init ref="id0" />
21. <transition>
22.   <source ref="id0" />
23.   <target ref="id1" />
24.   <label kind="guard">Aprodcons_1_buf == buf_put</label>
25.   <label kind="assignment">
26.         Aprodcons_1_buf = none,
27.         buf_Aprodcons_1_Sprodcons_1 = Sprodcons_ok_put
28.    </label>
29. </transition>
30. <transition>
31.   <source ref="id1" />
```

```
32.   <target ref="id2" />
33.   <label kind="guard">Aprodcons_1_buf == buf_put</label>
34.  <label kind="assignment">
35.         Aprodcons_1_buf = none,
36.         buf_Aprodcons_1_Sprodcons_1 = Sprodcons_ok_put
37.   </label>
38. </transition>
39. <transition>
40.   <source ref="id2" />
41.   <target ref="id1" />
42.   <label kind="guard">Aprodcons_1_buf == buf_get</label>
43.   <label kind="assignment">
44.         Aprodcons_1_buf = none,
45.         buf_Aprodcons_1_Sprodcons_1 = Sprodcons_ok_get
46.   </label>
47. </transition>
48. <transition>
49.   <source ref="id1" />
50.   <target ref="id0" />
51.   <label kind="guard">Aprodcons_1_buf == buf_get</label>
52.   <label kind="assignment">
53.         Aprodcons_1_buf = none,
54.         buf_Aprodcons_1_Sprodcons_1 = Sprodcons_ok_get
55.   </label>
56. </transition>
57. <transition>
58.   <source ref="id0" />
59.   <target ref="id1" />
60.   <label kind="guard">Aprodcons_2_buf == buf_put</label>
61.  <label kind="assignment">
62.         Aprodcons_2_buf = none,
63.         buf_Aprodcons_2_Sprodcons_2 = Sprodcons_ok_put
64.   </label>
65. </transition>
66. <transition>
67.   <source ref="id1" />
68.   <target ref="id2" />
69.   <label kind="guard">Aprodcons_2_buf == buf_put</label>
70.   <label kind="assignment">
71.         Aprodcons_2_buf = none,
72.         buf_Aprodcons_2_Sprodcons_2 = Sprodcons_ok_put
73.   </label>
74. </transition>
75. <transition>
76.   <source ref="id2" />
```

```
77.   <target ref="id1" />
78.   <label kind="guard">Aprodcons_2_buf == buf_get</label>
79.   <label kind="assignment">
80.         Aprodcons_2_buf = none,
81.         buf_Aprodcons_2_Sprodcons_2 = Sprodcons_ok_get
82.   </label>
83. </transition>
84. <transition>
85.   <source ref="id1" />
86.   <target ref="id0" />
87.   <label kind="guard">Aprodcons_2_buf == buf_get</label>
88.   <label kind="assignment">
89.         Aprodcons_2_buf = none,
90.         buf_Aprodcons_2_Sprodcons_2 = Sprodcons_ok_get
91.   </label>
92. </transition>
93. </template>
```

The declaration of the channels with initial messages (lines 1–3) and automata instances (lines 5–9) follows. Finally, the system is declared as a set of automata implementing the servers (line 10):

```
1. buf_service Aprodcons_1_buf = none;
2. buf_service Aprodcons_2_buf = none;
3. S_service Aprodcons_1_Sprodcons_1 = S_doSth;
4 //next channels...

5. buf_ = buf(1, Aprodcons_1_Sprodcons_1, Aprodcons_2_Sprodcons_2,
           Aprodcons_1_buf, Aprodcons_2_buf);
6. Sprodcons_1 = S(1, Aprodcons_1_put_butler, Aprodcons_1_get_butler, Aprodcons_1_buf,
           Aprodcons_1_Sprodcons_1);
7. Sprodcons_2 = S(2, Aprodcons_2_put_butler, Aprodcons_2_get_butler, Aprodcons_2_buf,
           Aprodcons_2_Sprodcons_2);
8. get_butler = butler(1, Aprodcons_1_Sprodcons_1, Aprodcons_2_Sprodcons_2,
           Aprodcons_1_get_butler, Aprodcons_2_get_butler);
9. put_butler = butler(2, Aprodcons_1_Sprodcons_1, Aprodcons_2_Sprodcons_2,
           Aprodcons_1_put_butler, Aprodcons_2_put_butler);

10. system buf_, Sprodcons_1, Sprodcons_2, get_butler, put_butler;
```

The temporal formula in Uppaal version is (the operator `-->` stands for **AG** ($p \Rightarrow$ **AF** q)):

```
Aprodcons_1_buf != none --> Aprodcons_1_buf == none
```

The Uppaal documentation and literature do not refer to fairness in verification. Indeed, in verification Uppaal reports a deadlock and gives the counterexample similar to Spin (Fig. 9.3). The loop in which *Aprodcons_1* repeatedly tries to *get* (*Sprodcons_1* switches to *cons*) while *Aprodcons_2* performs *get* and

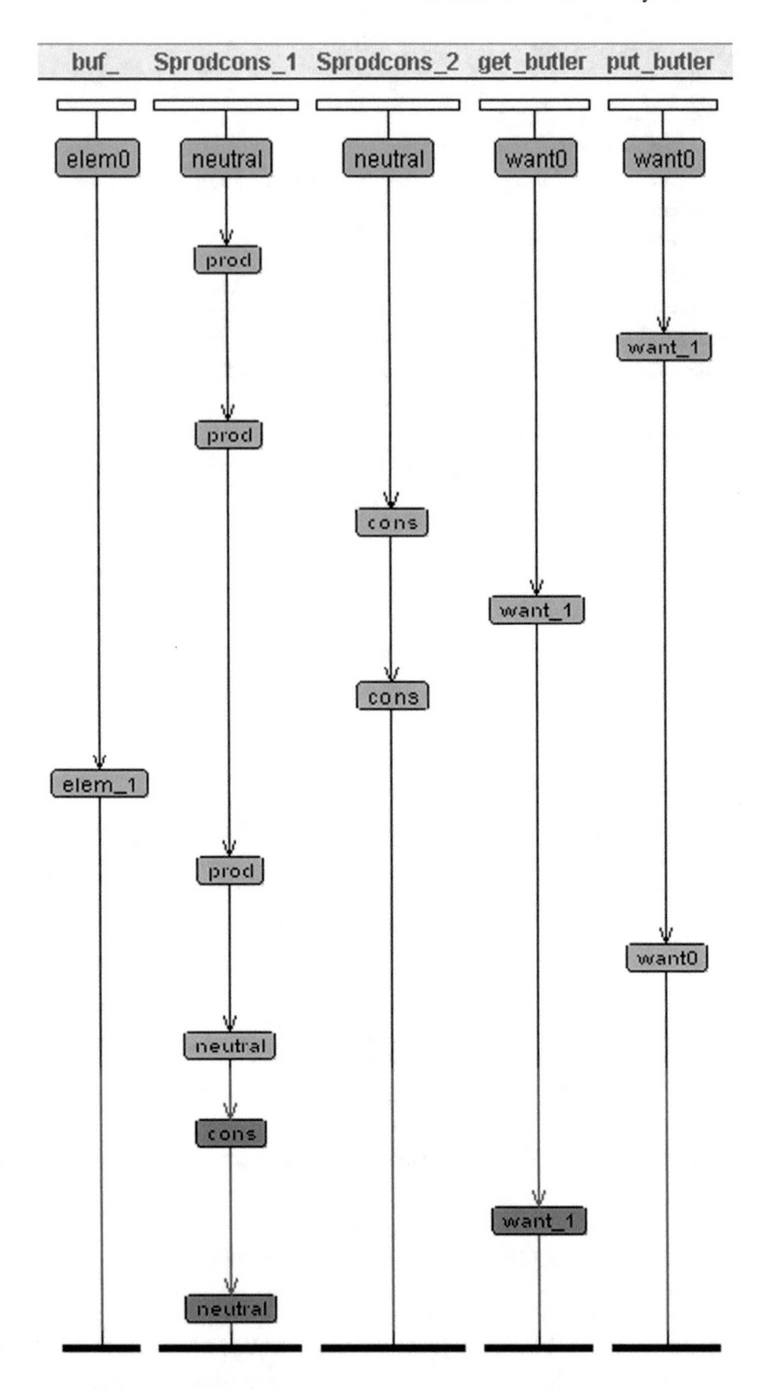

Fig. 9.3 Trace of the false deadlock identified by Uppaal in bounded buffer model

does not move, is highlighted by red locations on red arrows. Surprisingly, the simulation of the model under Uppaal shows normal operation, in which the variable `A_1_buf` changes to `put` or `get` and back to `none`.

9.5 Fair Verification Algorithm

The experience with commonly used verifiers: Spin, NuSMV and Uppaal shows that the verification of models with non-busy form of waiting gives proper results. However, problems with fairness in the case of busy form of waiting should be defeated. There are algorithms for LTL and CTL model checking with strong fairness (Kesten et al. 1998; Pnueli and Sa'ar 2008; Rozier 2011). Model checking with strong fairness requires the explicit specification of the events, which should be treated as compassionate. In the monograph, we present our own algorithm for CTL model checking, which supports compassion without specifying compassion points. In Dedan only a limited set of temporal formulas is needed for deadlock and termination detection, they are listed in Chap. 4. The algorithm is based on the "Checking by Spheres" algorithm (CBS (Daszczuk 2001, 2003)), with some improvements. The most important is using reverse reachability for discovering the shape of the reachability space. The basic CBS algorithm is a breadth-first search, constructing consecutive *spheres* of nodes of the reachability space, distant from the initial node for 1, 2, …, etc. transitions. The two finishing conditions are defined:

- Construction of an empty sphere (all nodes visited).
- A node with a given condition fulfilled in a non-empty sphere.

In (Daszczuk 2003), a temporal verifier TempoRG for the entire CTL is described, with some extensions for state quantification and component-aware operators. In order to verify in Dedan, the CBS algorithm was reduced to the two formulas necessary for the IMDS deadlock and termination detection: **EF AG** φ and **AF** φ. First, a decision is made if the reachability space is purely cyclic, or alternatively if there are nodes outside the cyclic part. In cyclic space, the fairness causes all execution paths to visit each node in the space infinitely often. In the other case, fairness causes each path to settle down in a strongly connected subgraph without escaping from it (we call it *ending strongly connected subgraph*—or *ending subgraph* in short, it is a generalization of the *lasso-shaped* ending subpath in (Rozier 2011)).

The verification of **EF AG** φ is based on reverse reachability, similar to (Hung and Chen 2006), illustrated in Fig. 9.4. The formula φ is $(D_s \wedge \neg E_s)$—communication deadlock or $(D_a \wedge \neg E_a)$—resource deadlock. The larger "cloud" is a cyclic part, the smaller clouds are ending subgraphs. The chessboard-filled circle is the initial node. First, all nodes fulfilling φ are identified (circles with dashed filling in Fig. 9.4a). Then a past of these nodes is found using reverse reachability (dotted filling of the ending subgraphs and the cyclic part in Fig. 9.4b). If any ending subgraph is left (subgraph with grilled filling in Fig. 9.4b), **EF AG** φ is *true*—a deadlock occurs.

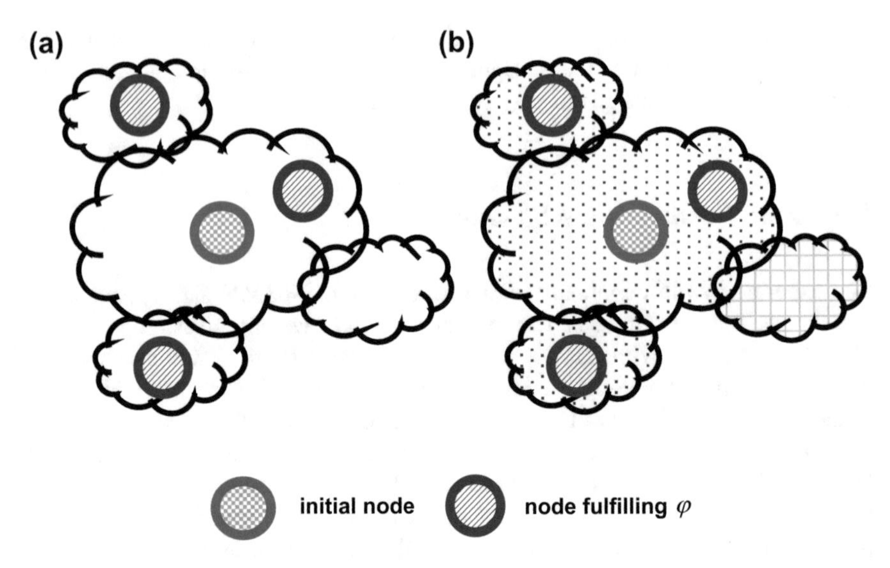

Fig. 9.4 Illustration of the verification of EF AG φ formula: **a** finding nodes fulfilling φ, **b** calculating the past of the nodes fulfilling φ

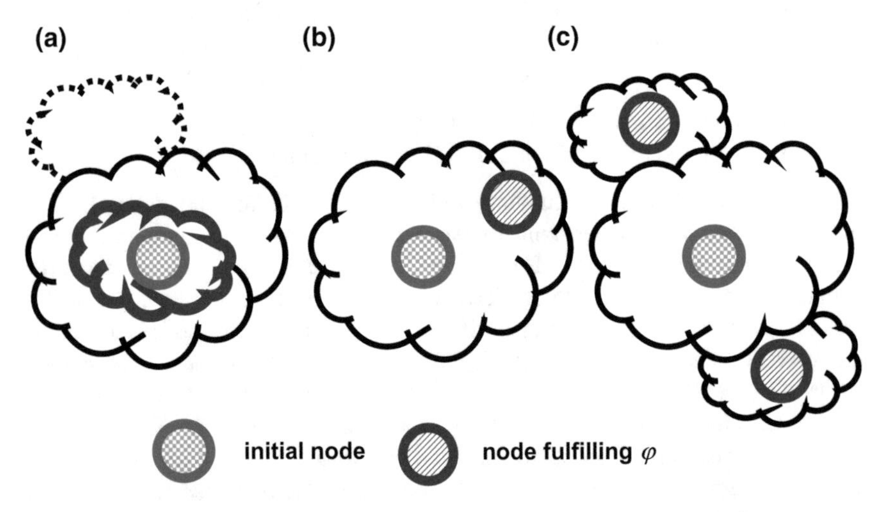

Fig. 9.5 Illustration of the verification of AF φ formula: case **a** the border of nodes fulfilling φ surrounding the initial node, case **b** cyclic space with a node fulfilling φ, case **c** ending strongly connected subgraphs, every one with a node fulfilling φ

The formula **AF** φ (where φ is F_a) is a termination condition—it is verified in three cases (Fig. 9.5). Again—the larger cloud is a cyclic part, the smaller ones are ending subgraphs (the edge of a "cloud" in Fig. 9.5a is dotted, because the ending subgraph can exist or not for this case). The initial node is a circle with a chessboard filling.

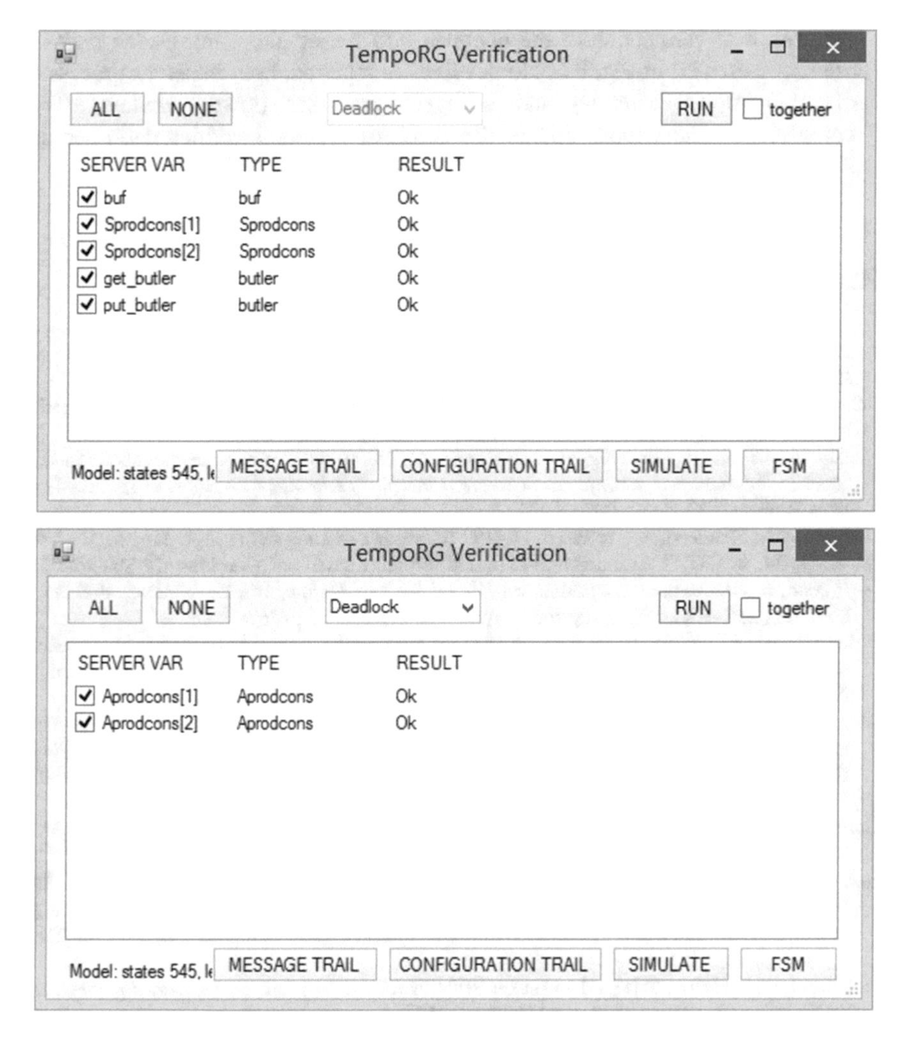

Fig. 9.6 Verification of bounded buffer example, using Dedan built-in verifier, server processes (upper) and agent processes (lower)

- The graph is cyclical or not, but there is a border surrounding the initial node, with each node satisfying φ (Fig. 9.5a). Compassion forces every path to cross the border—the formula is *true*.
- The graph is purely cyclical, and there is at least one node satisfying φ (Fig. 9.5b). Compassion forces every path to achieve this node—the formula is *true*.
- The graph contains ending subgraphs—compassion causes every path to fall into one of these subgraphs, and then to visit every node in this subgraph infinitely often. If there exists a state fulfilling φ in every ending subgraph (Fig. 9.5c)—the formula is *true*.

The result of verification of the bounded buffer example, using Dedan built-in verifier, is presented in Fig. 9.6. No deadlock is reported, both in server processes verification (upper window) and in agent processes (lower window). The "TempoRG" in the widow title is the name of temporal verifier described in (Daszczuk 2001, 2003).

References

Baier, C., & Katoen, J.-P. (2008). *Principles of model checking*. Cambridge, MA: MIT Press. ISBN: 9780262026499.

Behrmann, G., David, A., & Larsen, K. G. (2006). *A Tutorial on Uppaal 4.0*. Aalborg, Denmark. url: http://www.it.uu.se/research/group/darts/papers/texts/new-tutorial.pdf.

Cimatti, A., Clarke, E., Giunchiglia, F., & Roveri, M. (2000). NUSMV: A new symbolic model checker. *International Journal on Software Tools for Technology Transfer*, *2*(4), 410–425. https://doi.org/10.1007/s100090050046.

Clarke, E. M., Grumberg, O., & Peled, D. (1999). *Model checking*. Cambridge, MA: MIT Press.

Daszczuk, W. B. (2001). Evaluation of temporal formulas based on "Checking By Spheres." In *Proceedings Euromicro Symposium on Digital Systems Design, Warsaw, Poland, 4–6 Sept. 2001* (pp. 158–164). IEEE. https://doi.org/10.1109/dsd.2001.952267.

Daszczuk, W. B. (2003). *Verification of temporal properties in concurrent systems*. Warsaw University of Technology. url: https://repo.pw.edu.pl/docstore/download/WEiTI-0b7425b5-2375-417b-b0fa-b1f61aed0623/Daszczuk.pdf.

Dick, G., & Yao, X. (2014). Model representation and cooperative coevolution for finite-state machine evolution. In *2014 IEEE Congress on Evolutionary Computation (CEC), Beijing, China, 6–11 July 2014* (pp. 2700–2707). New York, NY: IEEE. https://doi.org/10.1109/cec.2014.6900622.

Gómez, R., & Bowman, H. (2005). *Discrete Timed Automata. Technical Report 3-05-2005*. Canterbury, UK. url: https://kar.kent.ac.uk/14362/1/TR305.pdf.

Holzmann, G. J. (1995). Tutorial: Proving properties of concurrent systems with SPIN. In *6th International Conference on Concurrency Theory, CONCUR'95, Philadelphia, PA, 21–24 Aug 1995* (pp. 453–455). Berlin Heidelberg: Springer-Verlag. https://doi.org/10.1007/3-540-60218-6_34.

Holzmann, G. J. (1997). The model checker SPIN. *IEEE Transactions on Software Engineering*, *23*(5), 279–295. https://doi.org/10.1109/32.588521.

Hung, Y. -C., & Chen, G. -H. (2006). Reverse reachability analysis: A new technique for deadlock detection on communicating finite state machines. *Software: Practice and Experience*, *23*(9), 965–979. https://doi.org/10.1002/spe.4380230904.

Katoen, J.-P. (1999). *Concepts, algorithms, and tools for model checking*. Germany: Erlangen-Nürnberg. url: http://www.cs.aau.dk/~kgl/VERIFICATION99/katoen2.ps

Kesten, Y., Pnueli, A., & Raviv, L. (1998). Algorithmic verification of linear temporal logic specifications. In *25th International Colloquium, ICALP'98 Aalborg, Denmark, 13–17 July 1998* (pp. 1–16). Berlin Heidelberg: Springer-Verlag. https://doi.org/10.1007/bfb0055036.

Laskowski, E., Tudruj, M., Olejnik, R., & Toursel, B. (2005). Java byte code scheduling based on the most-often-used-paths in programs with branches. In *The 4th International Symposium on Parallel and Distributed Computing (ISPDC'05), Lille, France, 4–6 July 2005* (pp. 21–27). New York, NY: IEEE. https://doi.org/10.1109/ispdc.2005.31.

Pnueli, A., & Sa'ar, Y. (2008). All you need is compassion. In *9th International Conference on Verification, Model Checking, and Abstract Interpretation, VMCAI 2008, San Francisco, CA, 7–9 Jan. 2008* (pp. 233–247). Berlin Heidelberg: Springer-Verlag. https://doi.org/10.1007/978-3-540-78163-9_21.

Rozier, K. Y. (2011). Linear temporal logic symbolic model checking. *Computer Science Review, 5*(2), 163–203. https://doi.org/10.1016/j.cosrev.2010.06.002.

Scholten, C. S., & Dijkstra, E.W. (1982). A class of simple communication patterns. In *Selected Writings on Computing: A personal Perspective* (pp. 334–337). New York, NY: Springer. https://doi.org/10.1007/978-1-4612-5695-3_60.

Zerzelidis, A., & Wellings, A. J. (2005). Requirements for a real-time .NET framework. *ACM SIGPLAN Notices*, *40*(2), 41. https://doi.org/10.1145/1052659.1052666.

Chapter 10
Timed IMDS

10.1 Timed Automata

The behavior of a distributed system may depend on real time flow (as opposed to succession of events in discrete systems), because changes in the system may take some time. As a result, deadlock situations may depend on relations between periods of time associated with individual changes. Several formalisms were invented to express the behavior of systems with time constraints, the most popular two are timed Petri nets and Timed Automata (Bérard et al. 2005; Lime and Roux 2006; Cassez and Roux 2006; Popescu and Martinez Lastra 2010). Furthermore, temporal logics related to timed systems were elaborated: Real-Time CTL (RTCTL, with temporal operators attributed with time constraints (Emerson et al. 1992; Gluchowski 2009), Clocked CTL (CCTL, based of time intervals in Discrete Time Systems (Ruf and Kropf 2003), Quantitative CTL (QCTL, with unit-delay transitions [Frossl et al. 1996]), Discrete Time CTL for embedded systems (DTCTL (Krystosik and Turlej 2006)) and Timed CTL (TCTL, connected with Timed Automata (Alur and Dill 1994)).

Timed Automata (TA (Alur and Dill 1994)) are similar to Büchi automata (Holzmann 1995) with time constraints. The automata execute their transitions independently, (in interleaving manner (Winskel and Nielsen 1995)), with the exception of transitions on common *symbols* (also called *signals*) which synchronize the automata. A set of real-time clocks is introduced for the time constraints,. Clocks are variables with values from $\mathbb{R}_{\geq 0}$, measured in a *basic unit of time* (a *unit* in short). The value of each clock is a real number, but we can observe the clocks when their values are integers or between integers. For example, having two clocks x and y we can look at the system in a situation in which x is 2 and y is between 0 and 1. In addition, the relation between the clocks (as $x < y$ or $x < y + 1$) can be observed. Transitions of timed automata are supplemented by time restrictions, which define the possible values of clocks, when individual transition can be fired (for example, $x \geq 1$). Time invariants can also be imposed on states of automata

© Springer Nature Switzerland AG 2020

W. B. Daszczuk, *Integrated Model of Distributed Systems*, Studies in Computational Intelligence 817, https://doi.org/10.1007/978-3-030-12835-7_10

(called *locations* in TA). The invariant informs how long the automaton can stay in the location, expressed as the relation over clocks, like $y < 3$. The clocks can be reset on transitions.

The transitions of TA are executed instantly, the clocks are advanced while automata stay in their locations. Therefore, the two types of progress are possible in TA: executing the transitions and progress of time. A detailed description of Timed Automata and their semantics are given in Alur and Dill (1994), Bérard (2013).

The semantics of a set of Timed Automata is based on two types of transitions: progress transitions and timed transitions:

- The *progress transition* (or simply: *transition*, also called *action transition* (Bowman 2001)) is executed if the Boolean expression on the arc of the automaton is fulfilled. This expression is compound over clocks and integer constants. If two clocks are used in an expression, only a difference between them is allowed. If more than one transition can be executed (in the same or in distinct automata), the choice is nondeterministic. The transitions are attributed with symbols called *actions* (do not confuse with IMDS actions, we use the term *symbol* for TA action). If the transitions in two or more automata have the same symbol, they are executed synchronously. Otherwise, the transitions are executed in an interleaving manner.
- If all the automata in the set have their time invariants in their current locations fulfilled, and the clock values used in all the locations are less than the maximum values of their invariants, then a *timed transition* can be executed. The execution is based on advancing all clocks synchronously (by the same, real value > 0). The maximum value the clocks can be advanced is the minimum difference between a maximum value of a clock used in an invariant and the current value of a clock used in this invariant.

If both the progress transition and the timed transition can be executed, the choice in nondeterministic. Note that if the timed transition is enabled, possibly infinitely many similar timed transitions are enabled (see Fig. 10.1b). Therefore, timed *regions* (classes of equivalence) are introduced into the formalism; they are limited by integer values (the *maximum value* used for a given clock is the greatest integer with which the clock is compared) and equal sign of fractional part of clock differences (0 is treated as a separate, third sign). This creates a set of *regions* of values of the clocks. An exemplary set of regions is shown in Fig. 10.1a, for two clocks x and y and maximum integers 3 for x and 2 for y. The relations between the clocks are not analyzed above the maximum values.

Because infinitely many timed transitions are possible between a pair or regions (Fig. 10.1b), a graph of region succession is constructed.

The construction of TA has important consequences, which can be traps for a designer who creates only timeless specifications. First, the system without any time constraints ("*timeless*" system) cannot be treated as the corresponding *timed system* without time constraints. Having the formula φ fulfilled in the initial location, the formula **AG** ($\varphi \Rightarrow$ **AF** ψ) may be false due to staying in the initial

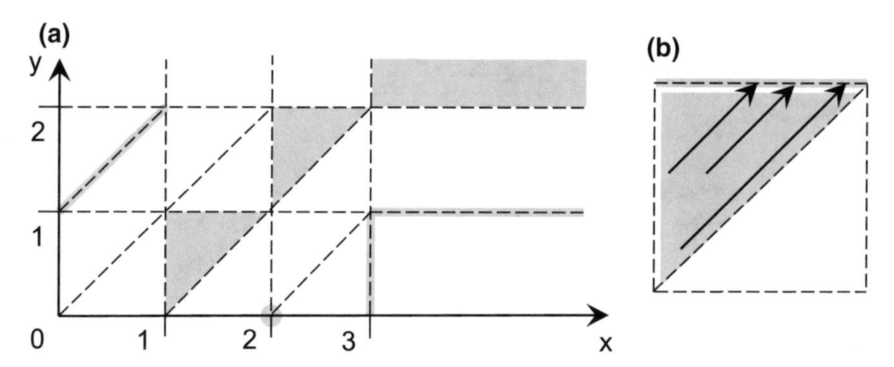

Fig. 10.1 **a** Example regions for two clocks x and y and maximum integers 3 for x and 2 for y. Diagonal: $0 < x < 1$, $1 < y < 2$, $x == y - 1$. Point: $x == 2$, $y == 0$. Vertical: $x == 3$, $0 < y < 1$. Horizontal: $x > 3$, $y == 1$. Upper triangle: $2 < x < 3$, $1 < y < 2$, $x - \lfloor x \rfloor < y - \lfloor y \rfloor$. Lower triangle: $1 < x < 2$, $0 < y < 1$, $x - \lfloor x \rfloor < y - \lfloor y \rfloor$. Rectangle: $x > 3$, $y > 2$. ($\lfloor x \rfloor$ stands for integer part of x, $x - \lfloor x \rfloor$ is fractional part). **b** Multiple timed transitions between a pair of regions (triangle and upper horizontal segment)

location forever, even if the Boolean condition for leaving the initial location is fulfilled. The lack of invariant in the initial location is equivalent to the invariant of the form $c \geq 0$, where c is a hypothetical clock starting from 0 in the initial location. Such a situation is illustrated in Fig. 10.2a. In the figure, invariants are shown inside the locations, time restrictions of transitions are given in regular font, and operations executed on transitions (as clock reset) are given in italics. Boolean conditions triggering the transitions are written in bold.

In order to convert a timeless system into a corresponding timed system, the following rules should be applied, illustrated in Fig. 10.2b (for the exemplary automaton *aut*):

- a local clock c_{aut} should be added;
- each location of the automaton *aut* should be equipped with the time invariant $c_{aut} = 0$;
- each transition in the automaton *aut* should reset the local clock c_{aut}.

In such a system, time does not flow and everything happens in zero time. This is appropriate, if we want to build a timeless system in Timed Automata. But if real time constraints are imposed on the system, timeless loops are dangerous, because they allow infinitely many events to happen in finite time. Such a situation is called *zero-time loop* (generalized as *zenoness*—infinitely many events in a finite time), and there are special tools to find them (Bowman et al. 2005; Hadjidj et al. 2007; Gómez and Bowman 2007). The example of a zero-time loop is given in Fig. 10.2c.

Another trap occurs when the location without time invariant has an outgoing transition with a maximum time bound, as in Fig. 10.2d. This location allows for staying in it forever, even if a Boolean condition allowing the escape is fulfilled during the period of the clock constraint of the transition. This can be cured by

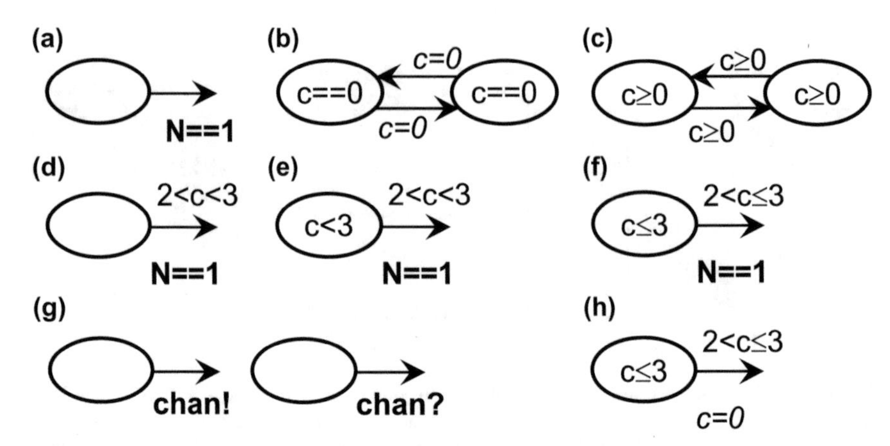

Fig. 10.2 Examples of locations and transitions in Timed Automaton: **a** location and transition without time constraints, **b** timed system corresponding to timeless system (instant invariants, clock resets on transitions), **c** zero-time loop, **d** location with infinite invariant, **e** location with finite invariant, open upper bound, **f** location with finite invariant, closed upper bound, **g** synchronous execution of transitions in two automata, synchronized by a channel (Uppaal TA) **h** resetting a clock on a transition. Condition in a state—time invariant. Regular font—time constraints of a transition. Italics font—update executed on a transition. *Bold font*—Boolean condition firing a transition or synchronization by channel

posing an invariant over the location: Fig. 10.2e shows the open upper bound while Fig. 10.2f shows the closed upper bound. However, this should be used carefully, because if the Boolean condition becomes fulfilled after the maximum time bound of an outgoing transition, it cannot be executed.

The original TA execute transitions synchronously if they are triggered by common symbols (which enable transitions). More than two automata can synchronize on a common symbol. In Uppaal, common symbols triggering transitions are replaced by communication channels. Communication through channels consists in sending and receiving a signal synchronously, for example *chan!* and *chan?*. This solution better models distributed systems, because the direction of communication is defined. Broadcast communication consists in receiving a signal sent from an automaton, by many automata.

The channel can be declared as *urgent*, in which case enabled transitions (with common channel symbols enabling these transitions: *chan!* or *chan?*) have a priority over time flow in locations. Therefore, a communication via urgent channel cannot be delayed if it is enabled. The transition enabled by urgent channel communication is also called urgent. The transition fired by the signal sent through the urgent channel has a priority over staying in the location. But the use of such a mechanism loses the asynchronous communication that is one of the key features of IMDS. Therefore, asynchronous communication via channels must be introduced separately.

Urgent transitions (with *chan!* or *chan?* as labels, Fig. 10.2g) cannot be mixed with expressions over clocks, as in Fig. 10.2c–f. Therefore, urgent channels and location invariants with expressions over clocks on transitions cannot be used together. For this reason, the two mechanisms of limiting the time spent in locations cannot be applied together.

10.1.1 Uppaal TA Syntax

Timed automata are usually defined as a set of timed automata and a set of real-valued clocks. For compatibility with Uppaal TA, we extend the definition of TA to cover the variables that can enable the transitions and can be assigned on transitions:

The Uppaal timed automaton *UTA* is a tuple

$$(L; l_0; Z; Lab; E; J_l, O, \{\bar{O}\}, \bar{O}_0) \quad (10.1)$$

where:

- $L = \{l_0, l_1, \ldots\}$ is the finite set of locations
- $l_0 \in L$ is the *initial location*
- $Z = \{c_0, c_1, \ldots\}$ is the set of *clocks*
- *Lab* is the set of actions labels (interpreted as actions on transitions), they are ($ch!$, $ch?$ and internal τ-actions)
- $J_l(l)$ maps every location $l \in L$ to the set of Z clock valuations over a Cartesian product of $\mathbb{R}^Z_{\geq 0}$, for example $J_l(l) = \{c_1 - c_2 > 2\}$
  Comment As in verification tools e.g. Uppaal (Lindahl et al. 2001), we restrict location invariants to constraints that are downwards closed, in the form: $x \leq n$ or $x < n$ where n is a natural number
- $O = \{o_1, o_2, \ldots\}$ the finite set of *variables*
- $W_i = \{w_{i1}, w_{i2}, \ldots\}$ the finite, integral set of *values* of variable $o_i \in O$
- $\bar{O} = (w_1, w_2, \ldots)$; $w_i \in W_i$—the *vector of values* of variables
- $\{\bar{O}\} = W_1 \times W_2 \times \cdots \times W_{card(O)}$—the *set of vectors* of variable values in O, $\bar{O} \in \{\bar{O}\}$
- $\bar{O}_0 = (w_1^0, w_2^0, \ldots)$—the initial vector of values of variables
- $E \subseteq L \times Lab \times J_{ll} \times 2^Z \times b \times F \times L$—the set of transitions
  - $J_{ll}(l, l')$ maps every transition $l, l' \in L$ to a set of Z clock valuations over a Cartesian product of $\mathbb{R}^Z_{\geq 0}$, just as $J_l(l)$ for a location l
  - 2^Z indicates the subset of clocks in Z that are reset on the transition
  - $b \subseteq \{\bar{O}\}$—the set of vectors of variable values enabling the transition, vectors containing a value r as xth element are denoted $b = \{\bar{O}[x] = r\}$
  - F: $\{\bar{O}\} \rightarrow \{\bar{O}\}$ the function assigning new values to the variables in O;

Comment In Uppaal TA, the function is given as assignment operation "variable: = expression over variables and integer constants" (all other variables are left unchanged).

10.1.2 UTA Semantics

The Semantics of the UTA is defined as follows:

Let

$$(L, l_0, Z, Lab, E, J_l, O, \{\bar{O}\}, \bar{O}_0) \text{ be an } UTA. \tag{10.2}$$

The semantics is defined as the Labeled Transition System LTS = $\langle Nodes, node_0, \succ \rangle$, where:

- $Node \subseteq L \times \{\bar{O}\} \times \mathbb{R}^Z$ is the set of nodes,
- $node_0 = (l_0, \bar{O}_0, u_0)$ is the initial node,
- $\succ = \succ_d \cup \succ_a$ is the transition relation such that
  - $(l, \bar{O}, u) \succ_d (l, \bar{O}, u + d)$ if $\forall_{d'}\ 0 \leq d' \leq \mathrm{d} \Rightarrow u + d' \in J_l(l)$
  - $(l, \bar{O}, u) \succ_a (l', \bar{O}', u')$ if there exists $e = (l, lab, g, r, b, F, l') \in E$ such that $u \in J_{ll}(l, l')$; $u' = [r \hookrightarrow 0]u$ and $\bar{O} \in b$ and $\bar{O}' = F(\bar{O})$

where:

- for $d \in \mathbb{R}_{\geq 0}$, $u + d'$ maps each clock x in Z to the value $u(x) + d'$,
- $[res \hookrightarrow 0]u$ denotes the clock valuation which maps each clock in *res* to 0 and agrees with u over $Z \backslash res$.

10.1.3 Network of UTA Syntax

The network of Uppaal Timed Automata over a common set of clocks and labels (actions), a common set of variables and a common initial vector of their values consists of n UTA:

$$UTA_i = (L_i, l_i^0, Z, Lab, E_i, J_{li}, O, \{\bar{O}\}, \bar{O}_0) \quad \text{with } 1 \leq i \leq n. \tag{10.3}$$

A location vector is a vector $\bar{Y} = (l_1, \ldots, l_n)$.

Location invariant functions are composed into a common function over location vectors $J_l(\bar{Y}) = J_{l1}(l_1) \wedge \ldots \wedge J_{ln}(l_n)$.

$\bar{Y}[l_i / l_i']$ denotes the location vector where the ith element l_i of $\bar{Y}$ has been replaced by l_i'.

O is the common set of variables, $\{\bar{O}\}$—the common set of vectors over O, $\bar{O}_0$—the common vector of their initial values.

10.1.4 Network of UTA Semantics

Let

$$UTA_i = (L_i, l_i^0, Z, Lab, E_i, J_{li}, O, \{\bar{O}\}, \bar{O}_0). \tag{10.4}$$

Let $\bar{Y}_0 = (l_1^0,..., l_n^0)$ be the *initial location vector*.

The semantics is defined as the Labeled Transition System LTS = $\langle Nodes, node_0, \succ \rangle$, where:

- $Nodes = (L_1 \times \cdots \times L_n) \times \{\bar{O}\} \times \mathbb{R}^Z$ is the set of nodes,
- $node_0 = (\bar{Y}_0, \bar{O}_0, u_0)$ is the initial node,
- $\succ = \succ_d \cup \succ_a \cup \succ_!$ is the transition relation defined by:
  - $(\bar{Y}, \bar{O}, u) \succ_d (\bar{Y}, \bar{O}, u + d)$ if $\forall_{d'}\ 0 \le d' \le \mathrm{d} \Rightarrow u + d' \in J_l(\bar{Y})$
  - $(\bar{Y}, \bar{O}, u) \succ_a (\bar{Y}[l_i/l_i'], \bar{O}', u')$ if there exists $(l_i, \tau; J_{ll}(l_i,l_i'), res, b_i, F_i, l_i') \in E_i$ such that $u \in J_{ll}(l_i,l_i')$; $u' = [res \hookrightarrow 0]u$ and $\bar{O} \in b_i$ and $\bar{O}' = f_i(\bar{O})$
  - $(\bar{Y}, \bar{O}, u) \succ_! (\bar{Y}[l_j/l_j', l_i/l_i'], \bar{O}', u')$ if there exist $(l_i, ch?, J_{ll}(l_i,l_i'), res_i, b_i, F_i, l_i') \in E_i$ and $(l_j, ch!, J_{ll}(l_j, l_j'), res_j, b_j, F_j, l_j') \in E_j$ such that $u \in J_{ll}(l_i, l_i') \cap J_{ll}(l_j, l_j')$, $u' = [res_i \cup res_j \hookrightarrow 0]u$ and:

    $(\bar{O} \in b_i$ and $\bar{O}' = F_i(\bar{O})$ and $\bar{O} \in b_j$ and $F_j(\bar{O}) \equiv \bar{O})$ or
    $(\bar{O} \in b_j$ and $\bar{O}' = F_j(\bar{O})$ and $\bar{O} \in b_i$ and $F_i(\bar{O}) \equiv \bar{O})$
    Comment We require that at least one of the functions F_i, F_j must be identically equal to its argument (it is ignored), and the other one is in effect (gives new values of the variables). This requirement prevents incoherent assignments to the same variable in automata UTA_j and UTA_k; the requirement is fulfilled by construction in translation of IMDS/Uppaal, as an assignment is applied in at most one automaton from the pair.
  - If both $\succ_d$ and $\succ_!$ are possible from given node $(\bar{Y}, \bar{O}, u)$, then $\succ_d$ is not included into LTS.
    Comment In such a way, Uppaal urgent channels are achieved; only such channels are applied in translation IMDS/Uppaal.

10.2 Timed Extensions to IMDS (T-IMDS)

The timed version of IMDS (T-IMDS) is a natural extension. In T-IMDS, all the mentioned features of IMDS are preserved, combined with model checking: communication duality, locality, autonomy, asynchrony and automated verification. In addition, time constraints are imposed on elements of a distributed system that can last in the time interval:

(a) time duration of the actions (fixed or subrange),
(b) limitation of time spent in the states.

The latter feature can be achieved in two ways:

- *asynchronous channels* with limited *time delay* (implemented using Uppaal urgent channels and urgent transitions)—CT-IMDS;
- applying *time bounds* to the states (BT-IMDS):
  - state *time bound*—maximum time spent in the state,
  - transition *time restriction*—clock values in which the transition can be fired,
  - *clock reset*.

These two ways of limiting time spent in the state are described separately. IMDS actions are attributed with *time duration* limits:

{agent.server.service, server.state} → *(time duration limits) {agent.server′.service′, server.state′}*

Time constraints specifying the action duration can have one of two forms:

- integer number (fixed time of the action)—angle brackets or parentheses can be used; examples: ⟨0⟩, (1);
- subrange—two comma-separated numbers; subrange can be open or closed in each limit—angle bracket denotes closed limit while parenthesis means open limit; examples: (0, 1), ⟨1, 2⟩, (0,2⟩.

A single number is a fixed duration time of the action, the subrange specifies a minimum and a maximum duration time; any value is possible between the bounds (including the bounds if the subrange is closed).

Limiting the duration of actions is not enough, because the control of time spent in states must supplement the specification of real-time behavior. Two kinds of such control are possible in T–IMDS:

- asynchronous channels between servers (CT–IMDS), with limited or unlimited time of transferring messages through the channel;.
- limiting the time spent in the states and restricting the moment of firing the actions (BT–IMDS), with the possibility of resetting the clock counting these times.

The implementation is based on the Uppaal verifier. It does not allow the use of two mechanisms together: the signal triggering the transition cannot be mixed with the expression over clocks. Therefore, two versions of T-IMDS are prepared, which cannot be mixed and are described separately.

To make order in defining individual time limitations, we use different terms:

- *time duration* refers to the duration of the action;
- *time delay* refers to the message transfer;
- *time bound* refers to the time spent in the state;
- *time restriction* refers to the moment of firing the action;
- *time constraints* refer to all of the above terms together.

10.2.1 Asynchronous Channels with Limited Time Delay (CT-IMDS)

The *time delay* of inter-server communication is specified using the *channels* command:

channels {(time delay), →server(time delay), server→server (time delay)}

A simple time delay is the default for all unspecified server pairs. If it is preceded by a server identifier, it concerns all messages transferred to this server. An index can be used for a server that is an element of a vector (*server[index]*) or to all elements of a vector (without index). Other possibilities are: all elements of a vector except specified (*server[-index]*), elements whose index modulo *number1* gives *number2* (*server[number2%number1]*), or mixed. If a server pair is specified, the time delay concerns communication only between this pair (the indexing described above may be applied to both servers). Each time delay may be fixed or a subrange, with open or closed limits.

The example of time delay definition is:

channels {(1,2), → sem<1,2>, proc[1] → sem[2]<1>), proc[-1] → sem[0%2] (2, 3>}

If the channels are asynchronous, but the message transfer is instantaneous, the command has the form:

channels;

10.2.2 Time Bounds of States (BT-IMDS)

Determining the duration of actions may be insufficient, because the time spent in individual states is undefined and can vary from 0 to any positive real number. To limit this time, *time bounds of states* are introduced. In fact, the bounds specify the values of an automaton's own clock that are possible while staying in the state. The definition of state bounds has the form:

bounds {?clk(time bound)state1, ?clk(time bound)state2, ...}

where *clk* is the keyword and denotes the private clock of the automaton. Time bound specifies the upper limit of stay in the state and can be open or closed (right parenthesis or right angle bracket, for example *?clk(1)state*, *?clk(2>state)*. Likewise, *time restriction* can be imposed on firing the action (clock values at which the transition can be fired, do not confuse with the action duration):

?clk(time restrictions){agent.server.service, server.state} → *{agent.server'.service', server.state'}*

The syntax and meaning of *time restrictions* are the same as for the action duration. The local automaton clock can be *reset* by placing the exclamation mark after *clk* (or after time limits of the action firing restriction):

?clk!{agent.server.service, server.state} → *{agent.server'.service', server.state'}*

In the action as: $\{a.s_1.r_1, s.p_1\} \rightarrow (x, y)\, \{a.s_2.r_2, s.p_2\}$, all instances of the server type S use the same lower and upper limits x and y. The same refers to the state bounds of the form *?clk(y)state* and for time restrictions of firing the action *?clk(x,y) action*. The time bounds can be appointed to all elements of the state vector or to individual elements of the state vector. To define time restrictions individually, conditional definitions are included in IMDS. They have the form *?number* or *?–number* preceding the time bounds for the state or placed before the action.

The condition *?number* refers to the server instance appointed by the *number*, while *?-number* refers to all instances except the *number*. If time bounds are defined for individual instances of the server type, a default value (without an instance number) must be defined as well.

Similarly, time delays of channels connecting the servers may concern individual target servers (→*(x, y)*), entire server vectors or non-vectors servers (→*serv (x, y)*), individual server vector elements (→ *serv[n](x, y)*) or server pairs ($serv_1 \rightarrow serv_2(x, y)$).

10.3 Conversion of T-IMDS to Uppaal Timed Automata

The Uppaal verifier (Behrmann et al. 2006) uses Timed Automata (Alur and Dill 1994) with several extensions:

- automata types (called *templates*) and variables,
- global and local (to individual automata) variables of integral types,
- global and local (to individual automata) clocks,
- in addition to the time restrictions, the relation on integral variables and integer constants can be used in Boolean expressions on transitions,
- formal and actual parameters of automata variables, passed by value or by reference,

- equal symbols on transitions are replaced by *channels*; synchronous execution of transitions with the same symbol is replaced by sending and receiving a signal through the channel (in this way, the communication direction is introduced),
- channels can be defined as *urgent*, which means that synchronization (simultaneous execution of sending and receiving of a signal over the channel) has priority over staying in locations.

The verified Timed IMDS system is converted to Uppaal Timed Automata. The distributed server automata S-DA^3, equivalent to the IMDS specification in the server view, are converted according to the following rules:

- the S-DA^3 automata type is converted to the Uppaal automata template;
- the template has formal parameters: its instance number *myNum* (index in the server vector) and integral variables that implement communication between the automaton and other automata;
- every automaton implementing a server in equipped with *hidden clock c*, which is used to count the time durations of actions;
- the states of S-DA^3 are converted to the locations of TA;
- the action (transition) of S-DA^3 (IMDS action) is converted to a transition-location-transition sequence; the location between the transitions is used to counting the time duration of the action;
- if various instances of the server type have different initial states, an additional initial location is added with the selection.

The location between the transitions is used to implement the action time duration; it is illustrated in Fig. 10.3a: the agent *a* action invoking the service *r1* is executed on the server *s1* in the state *p1*. The action lasts between *x* and *y* time units. It sends the message (*a*, *s2*, *r2*) of the agent *a* to another server *s2*, invoking its service *r2*; the server *s1* changes its state from *p1* to *p1'*. The additional location is equipped with time invariant $c < y$ (in the case of closed upper limit it would be $c \leq y$). The transition outgoing from this location has the Boolean guard $x < c < y$. The result is that the outgoing transition is followed not earlier than *x* units and not later than *y* units since entering the location (clock *c* is reset on entry to this location). Of course, *x* and *y* must be substituted by constants in the model, or *#DEFINE* may be used. The expressions over variables *a_si* on transitions are discussed in detail in the description of message passing below.

If various server variables of the common server type have different initial states, an additional location is introduced. The selection is made upon the value of the integer *myNum* parameter, passed to the timed automaton as the number of server type instance. On the basis of the value of this parameter, the transitions to the appropriate initial states are selected (Fig. 10.3b).

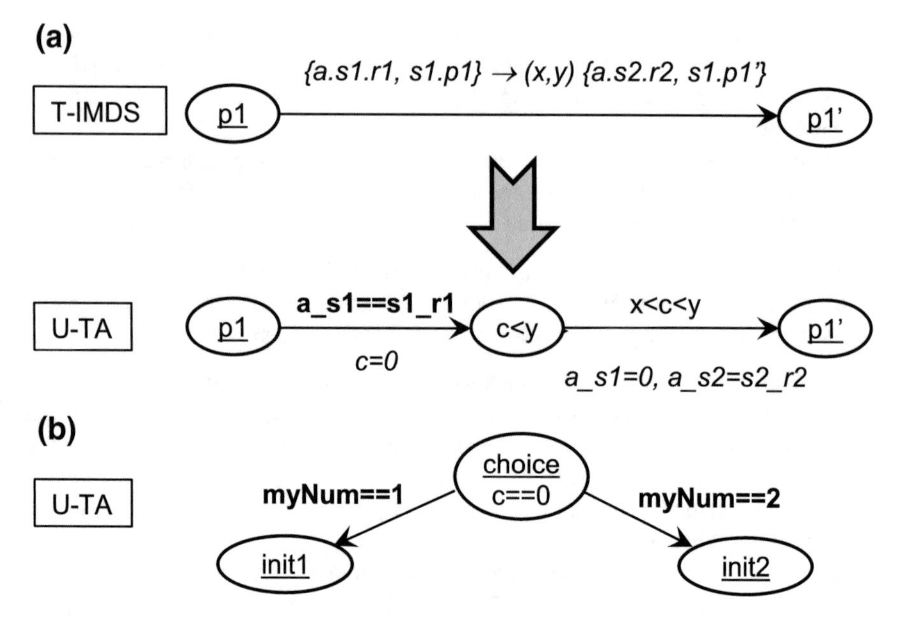

Fig. 10.3 Rules of conversion from T-IMDS to Uppaal TA: **a** transition of S-DA[3] implementing the action *{a.s1.r1, s1.p1} → (x,y) {a.s2.r2, s1.p1'}*, **b** Initial state dependent on automaton instance: `init1` for instance 1, `init2` for instance 2. Rules for Uppaal TA: Condition in a location—time invariant. Underlined font—state or location name. Regular font—time constraints of a transition. Italics font—update executed on a transition. Bold font—Boolean condition of firing a transition

10.3.1 Implementation of BT-IMDS

The implementation rules presented above are not enough, because the automaton can stay in a location infinitely, even if the condition firing the transition is fulfilled. Consider the Fig. 10.2a: we have the expression *N==1* as a transition guard, but the automaton can stay in the location infinitely even if *N* is really equal to 1, because there is no necessity to leave the state in TA (the timed transition can be chosen). The time spent in the state and the moment of leaving it can be determined by *state time bounds* and *action firing time restrictions*.

The BT-IMDS implementation is based on the following rules:

- if there are states with time bounds or actions with firing time restrictions specified (or explicit clock resets), an additional clock *clk* is added to count time flow for these bounds;
- the clock reset is implemented by updating $clk = 0$ on the transition;
- the state time bound is implemented by the location time invariant (over the clock *clk*);
- the action firing time restrictions are converted to the Boolean guard (over the clock *clk*) of the transition.

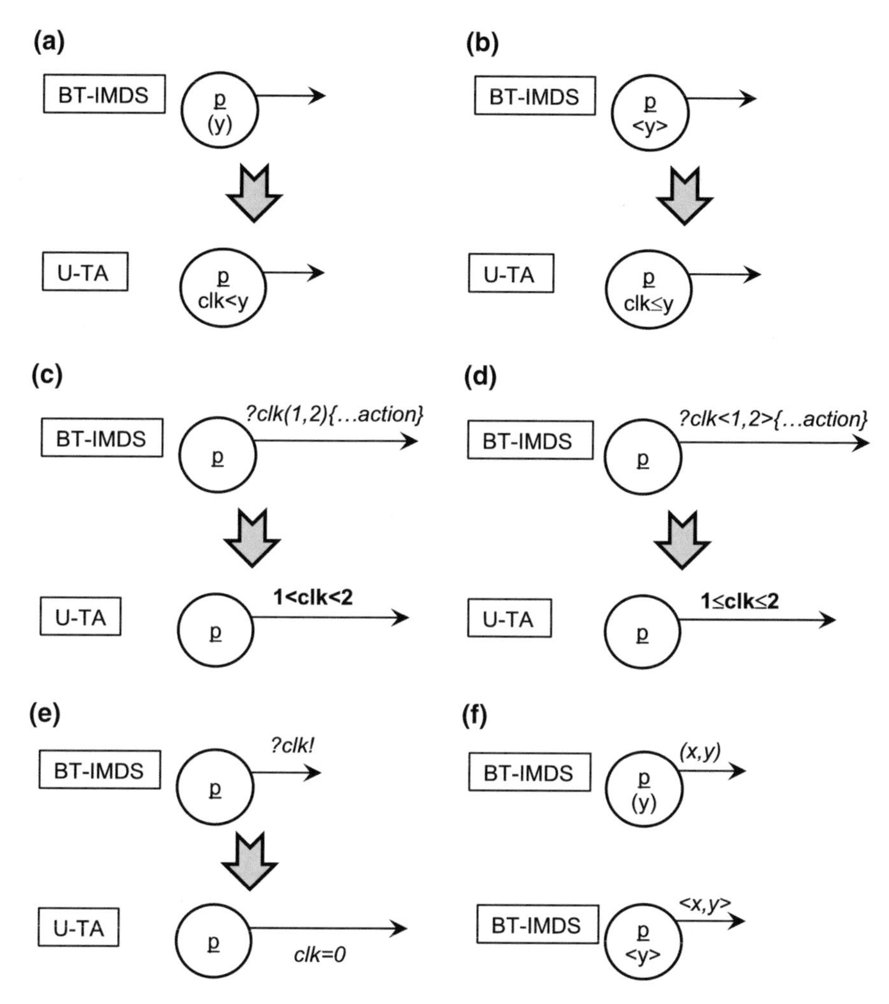

Fig. 10.4 Rules of conversion from T-IMDS with time bounds (BT-IMDS) of states to Uppaal TA (U-TA): **a** state with time bound, open limit, **b** state with time bound, closed limit, **c** time constraint of firing the action, open limits, **d** time constraint of firing the action, open limits, **e** clock reset in action, **f** coordination of state bound open/closed limit with open/closed time constraint upper limit

The state time bounds and action firing time restrictions are implemented using the internal automaton clock *clk*. The declaration *bounds* $p(y)$ causes the location p to use time invariant $clk < y$ (Fig. 10.4a). If the time bound is closed, the invariant has the form $clk \leq y$ (Fig. 10.4b).

Time restriction of firing the action *?clk(x, y){a.s.r, a.p}* → ... are implemented as Boolean guard over *clk* on the transition: $x < clk < y$, Fig. 10.4c. If the limits of time restriction are closed, the guard has the form $x \leq clk \leq y$ (Fig. 10.4d). Open and closed limits can be mixed: $(x, y\rangle$, $\langle x, y)$. The limit can be fixed: $\langle y, y\rangle$ or $\langle y\rangle$.

The clock *clk* is not reset automatically, therefore it may have a non-zero value on entering the state. Therefore, it can be explicitly reset on action leading to the state. The action resets *clk* if it is preceded by *?clk!* or *?(x, y)clk!* (Fig. 10.4e).

The value and kind (open or closed) of state time bound must be coordinated with the value and kind of upper limit of the firing restriction, as presented in Fig. 10.4f. Otherwise, if time bound is *bounds* $p\langle y\rangle$ and firing restriction is *?clk* (x, y), or if time bound is *bounds p(z)* and firing restriction is *?clk(x, y)*, $y < z$, then the state *p* cannot be left if the value of *clk* is *y* or greater than *y*, even if the action is prepared. This situation is sometimes called *time-lock* (Bowman 2001), a situation in which time is prevented from passing beyond a certain point. The zero-time loop (zeno loop (Hadjidj et al. 2007)) is an example of a time-lock, if there is no escape from the loop (in an unfair system, even an escape does not guarantee that the system leaves the loop). Other time-lock is presented in Fig. 10.2e, if the variable *N* does not get value 1 while the clock *c* value is $2 < c < 3$.

The state time bounds with action firing time restrictions should be used carefully, because if the condition preparing the action (the message arrives and starts to pend) is not fulfilled during the time limit specified in state bound, a time-lock occurs.

BT-IMDS message passing is implemented using global variables, passed to individual automata by parameters in their headings. The sending server sets the global variable to a value representing the service to be invoked. The message is pending at the target server simply keeping the assigned value of the variable, which is passed as the actual parameter to both servers: the sending one and the receiving one. The value is reset to 0 on firing an action invoked by the message in the target server. Because many messages can be pending at the server (one message for each agent), the variables used to store pending messages must be declared for each pair (*agent*, *target server*). This is illustrated in Fig. 10.5: the server *s1* in the action prepared in the state *p1* issues a message (*a*, *s2*, *r2*) to the server *s2*. The server *s2* can fire the action invoked by the message (*a*, *s2*, *r2*) in the state *p2* (however, it does not have to be the current state of *s2*). The message is pending until the server *s2* fires the action and thus resets the variable used to implement the message. In TA implementation, sending a message (*a*, *s2*, *r2*) consists in setting the value of variable *a_s2* to *s2_r2*. Firing the action in *s2* by accepting the message in the state *p2* is implemented by a Boolean guard *a_s2 ==s2_r2* of the transition. Execution of this transition resets *a_s2* to 0. Sending messages through values of global variables makes the communication—sending and accepting messages—really asynchronous.

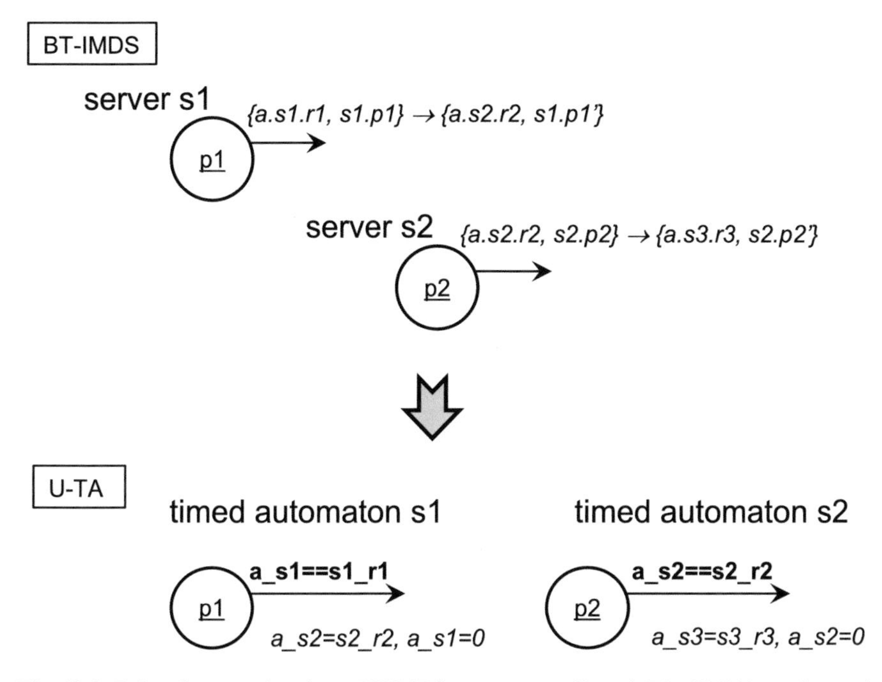

Fig. 10.5 Rule of conversion from BT-IMDS message to Uppaal TA (U-TA), setting and acceptance of a value of the variable implementing the communication channel

10.3.2 Implementation of CT-IMDS

The other way to enforce the firing of the prepared action in IMDS is to use *urgent channels* of Uppaal TA (synchronization on urgent channels has priority over staying in the location). Synchronization on urgent channels is called *urgent transitions*.

Synchronization in Timed Automata is based on synchronously executed transitions with common symbol in two automata. In Uppaal, the common symbol is equipped with *!* and *?* symbols, which denote *send* and *receive* operations on the channel. If the channel is defined as urgent, the send and receive operations have priority over time flow in locations (progress transitions have priority over timed transitions). The channels do not allow transferring the value from the sender to the receiver, therefore we call this type of communication *signals*. The message value (service identifier) must be transferred separately, by the variable value.

Boolean guards over clocks on urgent transitions are not allowed in Uppaal (Behrmann et al. 2006), therefore two versions of T-IMDS are elaborated: BT–IMDS includes state time bounds and action firing time restrictions (implemented using location time invariants and Boolean guards over clocks), while CT-IMDS includes asynchronous channels with time delays (implemented using Uppaal TA urgent channels).

The TA channels are synchronous, because the sender and the receiver simultaneously execute their transitions on the common channel. Therefore, urgent channels cannot be used directly to implement message passing in asynchronous IMDS. However, an asynchronous channel can be built using two synchronous channels: one for sending a message and the second for receiving it. Between sending and receiving operations, the time delay with specified limits can be applied. The second channel (the one used for receiving) can be used for keeping pending messages at the target server and for free choice of the message that matches the current state of the server.

The message passing implementation in Uppal TA is presented in Fig. 10.6. Global variables are used for transferring the message value, just as in BT-IMDS (or timeless IMDS converted to Uppaal TA). Setting the variable value is

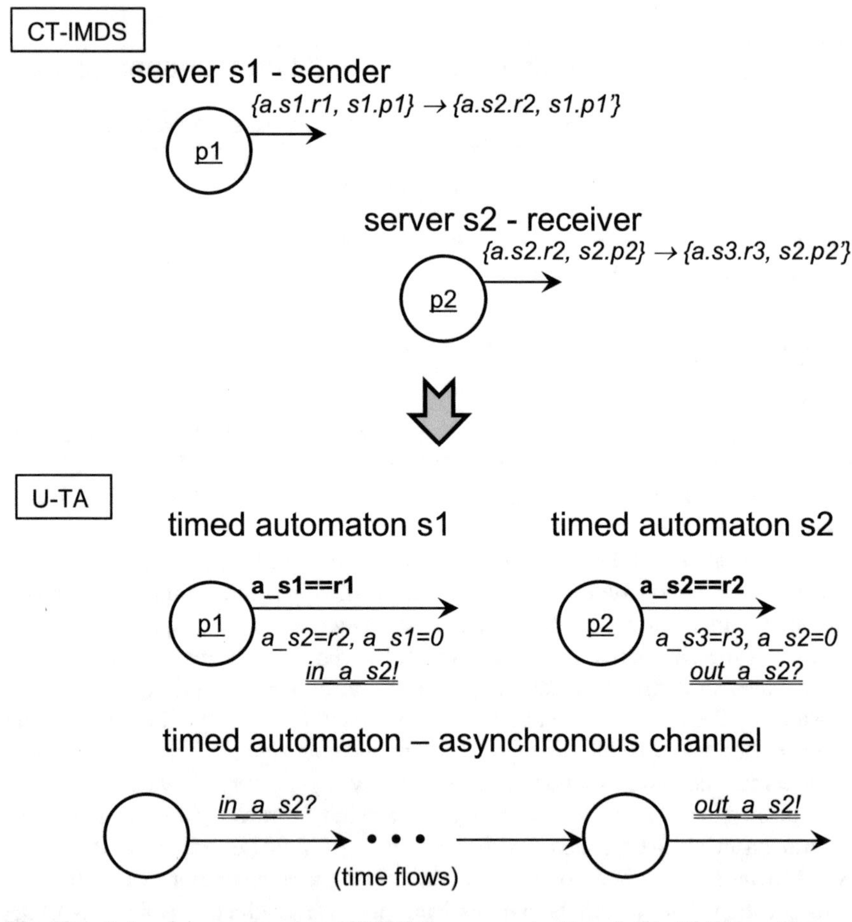

Fig. 10.6 Rule of conversion from CT-IMDS message to Uppaal TA (U-TA), setting and acceptance of a value and two synchronous channels conforming an asynchronous channel. Double underlined—synchronization on channels

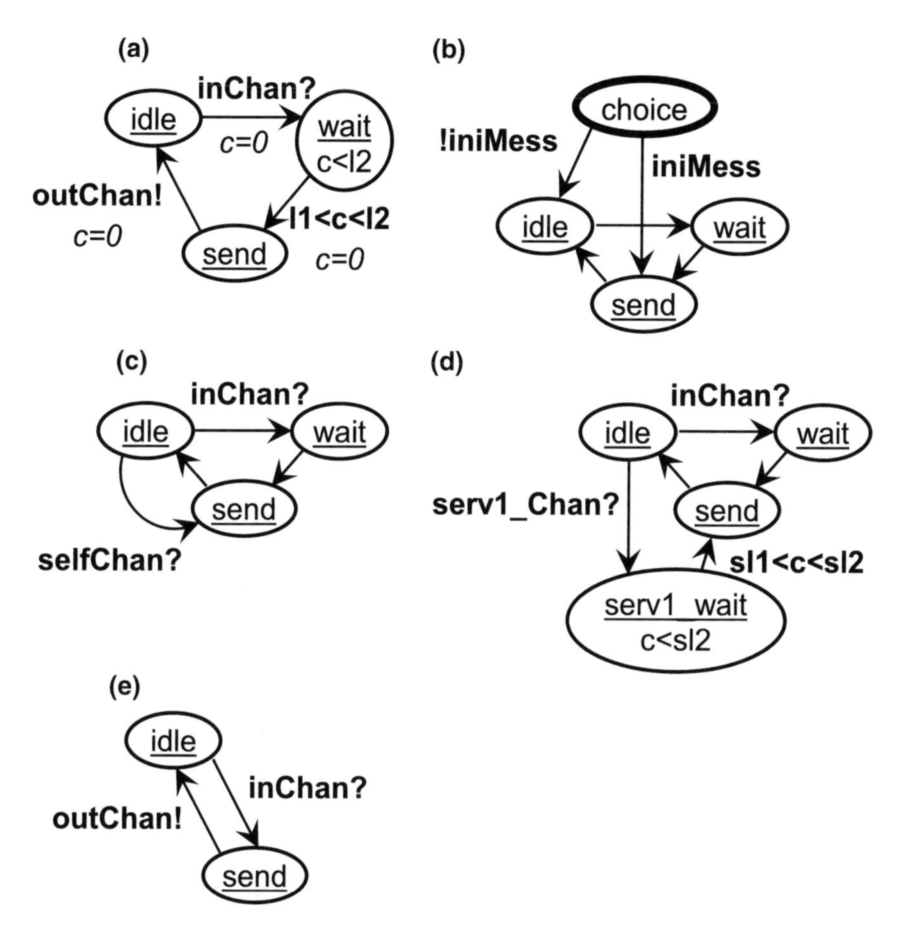

Fig. 10.7 Rules of conversion from T-IMDS to Uppaal TA, asynchronous channels: **a** basic asynchronous channel made of two synchronous channels *inChan* and *outChan*, **b** choice between existence of initial message and lack of it, **c** additional channel *selfChan* for instant self-sending a message, **d** time delay specific for source server *serv1*, **e** asynchronous channel with unspecified time delay

accompanied by sending a signal through an urgent channel. Execution of the transition enabled by the Boolean guard over the variable (implementing the message acceptance) is accompanied by receiving the signal over the second urgent channel. The two operations on two channels are coordinated in a separate timed automaton called *asynchronous channel*.

The basic structure of the asynchronous channel is presented in Fig. 10.7a. The inactive channel timed automaton stays in the *idle* location. Getting a signal from the sending automaton via *inChan* channel executes the transition to the *wait* location in which asynchronous channel stays for a period of time specified by

IMDS channel delay. The delay time is between lower and upper limit (*l1*, *l2*), specified in the IMDS *channels* command. Open and closed limits are allowed, as well as fixed delay $\langle l2, l2 \rangle$ or $\langle l2 \rangle$.

After the delay, the asynchronous channel moves to the *send* location. The transition to the *idle* location is executed synchronously with the transition in the receiver automaton using the *outChan* channel. Note that since the signal is received through *inChan*, the message value is kept in the global variable. It will be reset on acceptance of the message in the receiver automaton (see Fig. 10.6). Reaching the *idle* location finishes the cycle of transferring the message.

Asynchronous channels start in one of two ways. If the asynchronous channel initially contains the initial message of the agent, it starts in the *send* location. The value of the initial message (the service identifier) must be set to the global variable. Otherwise, the asynchronous channel starts in the *idle* location. The selection is done based the value of the Boolean parameter *iniMess*, which is observed on two outgoing transitions from the *choice* location and leading to the *send* and *idle* locations (Fig. 10.7b).

The message can be sent by the server to itself. This should be executed without time delay. For this purpose, the third urgent channel *selfChan* is used. Getting a signal through this channel leads directly to the *send* location, without intermediately staying in the *wait* location (Fig. 10.7c).

If a time delay is defined for the servers pair separately (*serv1* $\rightarrow$ *serv2*(*sl1*, *sl2*)), the additional input channel *serv1_chan* is defined for the server *serv1*, and a separate waiting location *serv1_wait* with its own time constraints is added (Fig. 10.7d).

The command *channels;* defines channels without a specified delay, therefore asynchronous channels without *wait* location are used (Fig. 10.7e).

Just like variables used to keep the message values, a channel must be defined for each pair (*agent*, *target server*).

The implementation of actions must follow the rules resulting from the construction of asynchronous channels: setting of a global variable which keeps the message value must be combined with sending a signal through the input channel of the asynchronous IMDS channel, while reading the global variable which keeps the message value must be combined with receiving the signal from the output channel of the asynchronous IMDS channel. This may be done safely if the action duration is 0 (Fig. 10.8a).

However, it is more complicated in the case of a non-zero duration. Recall that Boolean guards over clocks cannot be combined with synchronization over urgent channels in Uppal. Fortunately, this rule can be omitted with the help of a sequence of two transitions, as sending the message is performed at the end of the time-lasting action. It is illustrated in Fig. 10.8b: the location with the time invariant is passed first, then sending the signal follows.

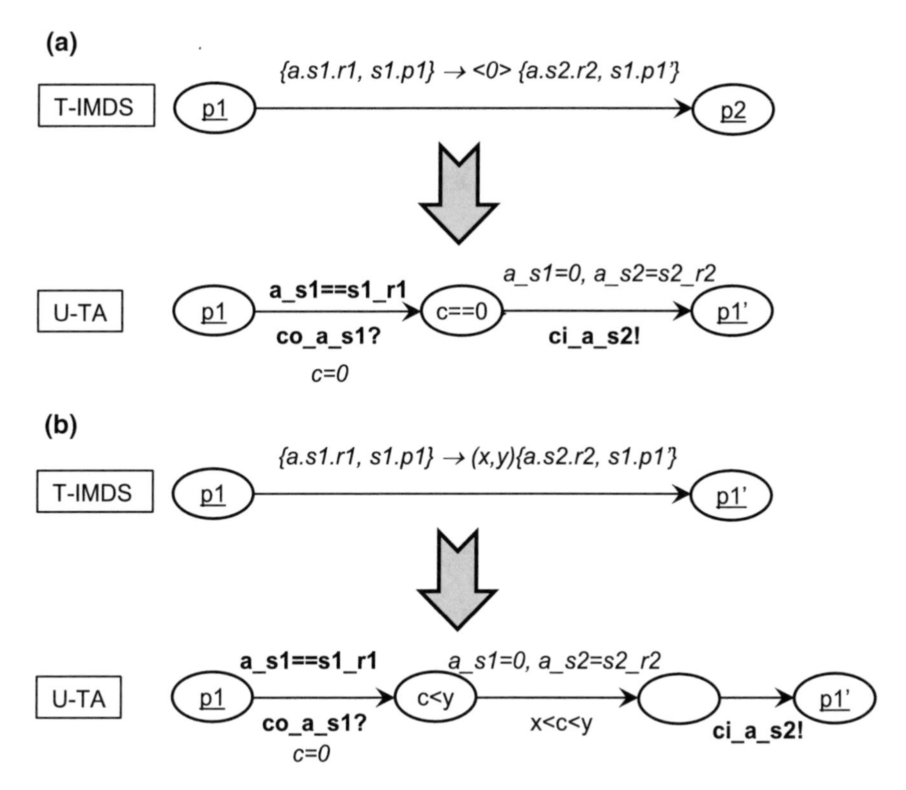

Fig. 10.8 Rules of conversion from T-IMDS to Uppaal TA (U-TA), time delays of message passing: **a** transition of S-DA[3] implementing an instant action (lasting for time 0), **b** transition of S-DA[3] implementing an action lasting for time between x and y

10.4 Formal Translation of CT-IMDS to Uppaal Timed Automata

The translation of CT-IMDS to Uppaal Timed Automata (UTA) should fulfil the following requirements:

- Every server is implemented as an UTA automaton, plus UTA implementing asynchronous channels.
- Every item is implemented as a set of locations of UTA and a set of UTA variables, plus UTA locations, UTA variables and UTA channels implementing asynchronous channels; therefore UTA global node represents the states of all automata, the messages of all non-terminated agents and phases of activity of asynchronous channels.
- Every action is implemented as a single UTA transition or a sequence of UTA transitions with no diverge until the action is completed (the input state that is left, the input message is consumed, the next state is reached, the next message is issued, and the action time duration is expired); the input items “disappeared” before counting time duration, the output items are established after it.

- Time constraints are imposed only on locations implementing the actions for counting action durations and on locations of UTA automata implementing asynchronous channels.

In this section, a translation of CT-IMDS to UTA is presented. A translation of BT-IMDS is analogous and is not covered.

10.4.1 Translation of Individual Servers to Uppaal Timed Automata

Every timed server $(s_j, B(s_j), g_{\Lambda sj})$ in the IMDS system of N servers and K agents is implemented as: $UTA = (L, l_0, Z, Lab, E, J, O, \{\bar{O}\}, \bar{O}_0)$

The ranges of agents 1,…,K and servers 1,…,N are used in shortened form 1..K and 1..N.

s_j—the IMDS server, $s_j \in S$.

$B(s_j)$—the transition relation of the IMDS server s_j (set of actions of a server s_j): $B(s_j) = \{\lambda \in \Lambda | \lambda = (((a,s_j,r),(s_j,p)),((a,s_k,r'),(s_j,p')))\}$

The set of constraints *constr* is a set of $\mathbb{R}_{\geq 0}$ subranges with natural bounds: $bool \times \mathbb{N} \times \mathbb{N} \times bool$, natural numbers are lower and upper bound of the subrange and Boolean values are kinds of lower and upper bounds (*false*-open/*true*-closed); the subrange must be monotonic, i.e., for (x_1, x_2, x_3, x_4), if $x_2 < x_3$ then both x_1 and x_4 can be of any kind (open or closed), while if $x_2 = x_3$ then $x_1 = x_4 = true$. The bounds $x_2 > x_3$ are prohibited. The notation is $x_1\ x_2, x_3\ x_4$, where open bounds x_1 and/or x_4 are denoted by round parentheses (), and while closed bounds x_1 and/or x_4 by angle brackets $\langle\ \rangle$. If $x_2 = x_3$ then single value can be supplied. Instant subrange is (*true*, 0, 0, *true*), denoted $\langle 0, 0\rangle$ or $\langle 0\rangle$. A special subrange is "unspecified", which means that the scope is $\mathbb{R}_{\geq 0}$. The notation of unspecified subrange is $\mathbb{R}_{\geq 0}$ or $\langle 0, \infty)$ or simply no subrange specified.

$Z = \{cs_j\}$—*hidden clock* of server s_j.

$g_{\Lambda sj}$: $B(s_j) \rightarrow constr$—the function mapping the actions of the server s_j to time durations.

$J_{\Lambda\ sj}(\lambda)$ is the action invariant mapping the clock cs_i to the subrange (*true*, 0, x_3, x_4) of $g_{\Lambda sj}(\lambda) = (x_1, x_2, x_3, x_4)$, $\lambda \in B(s_j)$.

L_{sj}—the set of locations $L_{sj} = P_{sj} = \{p_{sjx}(p_{sj}), p_{sj\lambda d}(p_{sj},\lambda, g_{\Lambda sj}(\lambda)), p_{sj\lambda y}(p_{sj},\lambda) | p_{sj} \in P(s_j), \lambda = ((m, p_{sj}),(m', p'_{sj})) \in \Lambda\}$ (three locations for every state p_{sj}). The set of locations connected with states of the server s_j; $p_{sjx}(p_{sj})$ represents the state p_{sj} itself (in which a message m can be accepted), $p_{sj\lambda d}(p_{sj}, \lambda, g_\Lambda(\lambda))$ is the location connected with the state p_{sj} and the action λ (more than one action can start from a given p_{sj}), used for counting the action λ duration, $p_{sj\lambda y}$ (p_{sj}, λ) is the location connected with the state p_{sj} and the action λ, in which the next message m' of the action λ is issued; the subsets of P_{sj} are

$P_{sjx} = \{p_{sjx}(p_{sj}) | p_{sj} \in P(s_j)\}$,
$P_{sjd} = \{p_{sj\lambda d}(p_{sj},\lambda,g_{\Lambda sj}(\lambda)) | p_{sj} \in P(s_j), \lambda = ((m, p_{sj}),(m', p'_{sj})) \in \Lambda\}$,
$P_{sjy} = \{p_{sj\lambda y}(p_{sj},\lambda) | p_{sj} \in P(s_j), \lambda = ((m, p_{sj}),(m',p'_{sj})) \in \Lambda\}$.

l_0—p_{sjx} of the initial state, $l_0 = p_{sjx}(p_{sj})$, $p_{sj} \in P_{sj} \cap P_{ini}$ (this intersection contains exactly one element).

$J(l)$ is the location time invariant; for timeless locations in $P_{sjx} \cup P_{sjy}$ of a server s_j, the invariant $J(l \in P_{sjx} \cup P_{sjy}) = \mathbb{R}_{\geq 0}$ for the clock cs_j. The invariant for timed locations P_{sjd}: $J(l \in P_{sjd}) = J_{\Lambda sj}(\lambda)$, $\lambda \in \Lambda$ for the clock cs_j (for actions on server s_j: $\lambda = (((a, s_j, r),(s_j, v)),((a,s_k, r'),(s_j, v')))$).

$Lab = B(s_j) \cup \{ch_{ij\_y\_k}!, ch_{ij\_x}?|i = 1..K, k = 1..N\}$—for server s_j: set of actions, labels of sending messages to servers $s_k \in S$ and labels of receiving messages.

$E \subseteq L \times g \times Lab \times \{ \{cs_j\}\} \times b \times F \times L = \{$
$(p_{sjx}(p_{sj}), \mathbb{R}_{\geq 0}, ch_{ij\_x}?, \{cs_j\}, b[i, j] = r, F(\bar{O}) \equiv \bar{O}, p_{sj\lambda d}(p_{sj}, \lambda, g_{\Lambda}(\lambda))),$
$(p_{sj\lambda d}(p_{sj},\lambda, g_{\Lambda sj}(\lambda)), g_{\Lambda sj}(\lambda), \lambda, \{cs_j\}, b[i,j] = r, F(\bar{O}) = \bar{O}/[i,k] = r', p_{sj\lambda y}(p_{sj}, \lambda)),$
$(p_{sj\lambda y}(p_{sj}, \lambda), \mathbb{R}_{\geq 0}, ch_{ij\_y\_k}!, \{cs_j\}, b[i, k] = r', F(\bar{O}) \equiv \bar{O}, p_{sjx}(p'_{sj}))\}$—

- for $\lambda = (((a_i, s_j, r),(s_j, v)),((a_i, s_k, r'),(s_j,v'))) \in \Lambda$; $\bar{O}/[i,k] = r'$ denotes that the result of F is $\bar{O}$ in which element $\bar{O}[i, k]$ is set to r' while all elements $\bar{O}[i, t] = novalue$, $t = 1..K$, $t \neq k$;
- for an action λ the sequence of transitions connecting the locations $p_{sjx}(p_{sj})$ and $p_{sjx}(p'_{sj})$ via the transition λ is: $(p_{sjx}(p_{sj}), p_{sj\lambda d}(p_{sj}, \lambda, g_{\Lambda}(\lambda)))$ labeled with $ch_{ij\_x}?$ (receiving of a message from channel ch_{ij_x}), transition $(p_{sj\lambda d}(p_{sj}, \lambda,g_{\Lambda}(\lambda)), p_{sj\lambda y}(p_{sj}, \lambda)))$ with time constraint $g_{\Lambda sj}(\lambda)$ and label λ, and transition $(p_{sj\lambda y}(p_{sj}, \lambda), p_{sjx}(p'_{sj}, \lambda)))$ labeled with $ch_{ij\_y\_k}!$ (sending the next message). Every transition resets the server s_j hidden clock cs_j.

The location $p_{sj\lambda d}(p_{sj}, \lambda, g_{\Lambda}(\lambda))$ has the time invariant $J_{\Lambda sj}(\lambda)$, all other locations have time invariant $\mathbb{R}_{\geq 0}$.

$O = \{o_{ij}|i = 1..K, j = 1..N\}$—the set of variables containing services called by the agents. Sending the message (a_i, s_j, r) is implemented as setting the value of the variable o_{ij} and simultaneous executing the *send* operation on the corresponding channel $ch_{ik\_y\_j}!$. Accepting the message (execution of the action with the message on input) is simultaneous resetting the variable to the neutral value (outside the set of services R, see below) and executing the receive operation on corresponding channel $ch_{ij\_x}?$.

$W_j = \{w_{1j}, w_{2j},\ldots\}$, $w_{ij} \in R \cup \{novalue\}$, $i = 1..card(R)$—the set of values implementing the set of services of the server s_j. In practice, only services used in messages directed to s_j can be used.

$\bar{O} = (w_{ij},\ldots)|i = 1..K, j = 1..N$; $w_{ij} \in W_j \cup \{novalue\}$—the vector of values of variables implementing messages ($\bar{O}[i, j] = w_{ij}$ is the service in the messages of the agent a_i pending at the server s_j). if $\bar{O}[i, j] \neq novalue$ then the message of a_i is pending at s_j, invoking the service $r = w_{ij}$. If $\bar{O}[i, k] = novalue$ for all $k = 1..N$ then the agent a_i is terminated.

$\bar{O}_0 = (w_{ij0},...)|i = 1..K,\ j = 1..N;\ w_{ij} \in W_j \cup \{novalue\}$—the initial vector of services of pending messages; if $\bar{O}[i, j] \neq novalue$ then the initial message of a_i is pending at s_j.

$\{\bar{O}\} = W_1 \times W_2 \times \cdots \times W_K$ is the set of vectors where $\bar{O} \in W_1 \times W_2 \times \cdots \times W_K$.

10.4.2 Semantics of Server Timed Automaton

The semantics of UTA is defined as a Labeled Transition System

$$\mathrm{LTS_a} = \langle Nodes;\ node_0;\ \succ\rangle, \tag{10.5}$$

where

- $Nodes \subseteq L \times \{\bar{O}\} \times \mathbb{R}^Z$ is the set of nodes,
- $node_0 = (l_0;\ \bar{O}_0;\ u_0)$ is the initial node,
- $\succ = \succ_d \cup \succ_? \cup \succ_\lambda \cup \succ_!$ is the transition relation such that:
  - $(l, \bar{O}, u) \succ_d (l, \bar{O}, u + d)$ if $\forall_{d'}\ 0 \leq d' \leq \mathrm{d} \Rightarrow u + d' \in J(l)$
  - $(p_{sjx}(p_j), \bar{O}, u) \succ_? (p_{sj\lambda d}(p_{sj}, \lambda, g_\Lambda(\lambda)), \bar{O}, u')$ if there exists $\lambda = ((m, p_{sjx}(p_j)), (m', p_{sjx}(p'_j))) \in \Lambda$ such that $m = (a_i, s_j, r)$, $m' = (a_i, s_k, r')$, $\bar{O}[i, j] = r$, $u \in J(p_{sjx}(p_j))$, $u' = [\{cs_j\} \hookrightarrow 0]u$—acceptance of the incoming message; note that the transition $(p_{sjx}, p_{sj\lambda d})$ is labeled with $ch_{ij\_x}?$; server s_j automaton transition $(p_{sjx}(p_j), \mathbb{R}_{\geq 0}, ch_{ij\_x}?, \{cs_j\}, \{\bar{O}[i, j] = r\}, F(\bar{O}) \equiv \bar{O}, p_{sj\lambda d}(p_{sj},\lambda, g_{\Lambda sj}(\lambda)))$
  - $(p_{sj\lambda d}(p_{sj}, \lambda, g_\Lambda(\lambda)), \bar{O}, u) \succ_\lambda (p_{sj\lambda y}(p_{sj}, \lambda), \bar{O}', u')$ for $\lambda = ((m, p_{sjx}(p_j)), (m', p_{sjx}(p'_j))) \in \Lambda$, such that $m = (a_i, s_j, r)$, $m' = (a_i, s_k, r')$, $\bar{O}'[i, j] = novalue$, $\bar{O}'[i, k] = r'$, $u \in g_{\Lambda sj}(\lambda) = J(p_{sj\lambda d}(p_{sj},\lambda,g_\Lambda(\lambda)))$, $u' = [\{cs_j\} \hookrightarrow 0]u$—transition to the location sending the next message (labeled with λ); server s_j automaton transition $(p_{sj\lambda d}(p_{sj},\lambda,g_{\Lambda sj}(\lambda)), g_{\Lambda sj}(\lambda), \lambda, \{cs_j\}, \{\bar{O}\}, F(\bar{O}) = \bar{O}/[i, k] = r', p_{sj\lambda y}(p_{sj}, \lambda))$
  - $(p_{sj\lambda y}(p_{sj}, \lambda), \bar{O}, u) \succ_! (p_{sjx}(p'_j), \bar{O}, u')$ for $\lambda = ((m,p_{sjx}(p_j)),(m',p_{sjx}(p'_j))) \in \Lambda$ such that $m = (a_i, s_j, r)$, $m' = (a_i, s_k, r')$, $u \in J(p_{sj\lambda y}(p_{sj},\lambda))$, $u' = [\{cs_j\} \hookrightarrow 0]u$ —transition to the next state and sending the next message; note that the transitions $(p_{sj\lambda y}(p_{sj}, \lambda), p_{sjx}(p'_j))$ is labeled with $ch_{ij\_y\_k}!$; server s_j automaton transition $(p_{sj\lambda y}(p_{sj}, \lambda), \mathbb{R}_{\geq 0}, ch_{ij\_y\_k}!, \{cs_j\}, \{\bar{O}\}, F(\bar{O}) \equiv \bar{O}, p_{sjx}(p'_j))$

where:

- for $d \in \mathbb{R}_{\geq 0}$, $u + d'$ maps clock cs_j in Z to the value $u(x) + d$,
- $[\{cs_j\} \hookrightarrow 0]u$ denotes the clock valuation which maps server s_j hidden clock cs_j to 0 and agrees with u over $Z\backslash clk$.

10.4.3 Translation of Asynchronous Channels to Uppaal Timed Automaton

The timed server (s_j, $B(s_j)$, $g_{\Lambda sj}$) of IMDS system of N servers and K agents is equipped with the set of asynchronous channels for every agent, implemented as UTA = (L; l_0; Z; Lab; E; J, O, $\{\bar{O}\}$, $\bar{O}_0$), where O, $\{\bar{O}\}$, $\bar{O}_0$ are not used. We use the name *asynchronous channel* to differentiate the automaton from synchronous channels used to implement the communication. The terms *sending* and *receiving* we reserve for servers, we say that asynchronous channel *gets* a message from the source server and *puts* it to the target server.

s_j—the IMDS server, $s_j \in S$, a_i—the IMDS agent, $a_i \in A$.

$Z = \{cc_{ij}\}$—the clock of asynchronous channel of the server s_j for agent a_i.

g_{chj}: $S \to constr$—the function mapping asynchronous subchannels (see below) from every server s_k, $k = 1..N$ to the server s_j, to time delays. For the self-subchannel $g_{chjj} = \mathbb{R}_{\geq 0}$ or $g_{chj} = (true, 0, \infty, false)$.

$J_{chj}(s_k)$ is the subchannel time invariant mapping cc_{ij} to the subrange ($true$, 0, x_3, x_4) of $g_{chj}(s_k) = (x_1, x_2, x_3, x_4)$.

$J(l)$ is the location time invariant; for timeless locations ch_{ij_idle} and ch_{ij_send} of the asynchronous channel ch_{ij}, the invariant $J(l \in \{ch_{ij\_idle}$ and $ch_{ij\_send}\}) = \mathbb{R}_{\geq 0}$ for the clock cc_{ij}. The invariant for timed location $ch_{ij\_wait\_k}$ $J(ch_{ij\_wait\_k}) = J_{chj}(s_k)$ for the clock cc_{ij}.

$Lab = \{ch_{ij\_y\_k}?|k = 1..N\} \cup \{ch_{ij\_x}!\} \cup \{delay\}$—for server s_j: set of actions, labels of getting messages from servers $s_k \in S$, label for putting a message to the server s_j and the label for counting delay of the channel s_j.

$L_{chij} = CH_{ij} = \{ch_{ij\_idle}, ch_{ij\_send}\} \cup \{ch_{ij\_wait\_k} \mid k = 1..N\}$—the set of locations of asynchronous channel for sending the agent a_i messages to the server s_j; ch_{ij_idle} is used for waiting for the agent a_i messages, $ch_{ij\_wait\_k}$ is the subchannel location used for counting the asynchronous channel delay, specific for the pair (s_k, s_j), and the location ch_{ij_send} is for putting a message to the server s_j; The subsets of CH_{ij} are $CH_{ij\_idle} = \{ch_{ij\_idle}\}$, $CH_{ij\_wait} = \{ch_{ij\_wait\_k}|k = 1..N\}$, $CH_{ij\_send} = \{ch_{ij\_send}\}$.

If the initial message of the agent a_i is directed to s_j, then l_0 (the initial location) is ch_{ij_idle}, otherwise it is ch_{ij_send}.

$E \subseteq L \times g \times Lab \times \{\{cc_{ij}\}\} \times b \times F \times L = \{$
$(ch_{ij\_idle}, \mathbb{R}_{\geq 0}, ch_{ij\_y\_k}?, \{cc_{ij}\}, \{\bar{O}\}, F(\bar{O}) \equiv \bar{O}, ch_{ij\_wait\_k})$,
$(ch_{ij\_wait\_k}, g_{chj}$ (k), $delay$, $\{cc_{ij}\}, \{\bar{O}\}, F(\bar{O}) \equiv \bar{O}, ch_{ij\_send})$,
$(ch_{ij\_send}, \mathbb{R}_{\geq 0}, ch_{ij\_x}!, \{cc_{ij}\}, \{\bar{O}\}, F(\bar{O}) \equiv \bar{O}, ch_{ij\_idle})\}$—
for the message sent from the server s_k to the server s_j, the sequence of transitions connecting the locations ch_{ij_idle}, $ch_{ij\_wait\_k}$, ch_{ij_send} is: (ch_{ij_idle}, $ch_{ij\_wait\_k}$) labeled with $ch_{ik\_y\_j}$? (getting the message from the channel $ch_{ik\_y\_j}$), transition ($ch_{ij\_wait\_k}$, ch_{ij_send}) with time constraint $g_{chj}(k)$ and label *delay*, and transition (ch_{ij_send}, $ch_{i\text{-}j\_idle}$) labeled with ch_{ij_x}! (putting the message to the server s_j using the channel ch_{ij_x}). Every transition resets the channel ch_{ij} clock cc_{ij}.

The location $ch_{ij\_wait\_k}$ has the time invariant $J_{chj}(k, j)$, all other locations have time invariant $\mathbb{R}_{\geq 0}$.

$O = \{o_{ij} | i = 1..K, j = 1..N\}$—the set of variables, not used in the automaton.

$\bar{O} = (w_{ij},...) | i = 1..K, j = 1..N$—the vector of values of variables, not used in the automaton.

$\bar{O}_0 = (w_{ij0},...) | i = 1..K, j = 1..N$—the initial vector of values of variables, not used in the automaton.

$\{\bar{O}\} \subseteq W_1 \times W_2 \times \cdots \times W_K$ is the set of vectors where $\bar{O} \in W_1 \times W_2 \times \cdots \times W_K$

10.4.4 Semantics of Asynchronous Channel Automaton

The semantics is defined as the Labeled Transition System

$$LTS_{ch} = \langle Nodes; node_0; \succ \rangle, \tag{10.6}$$

where:

- $Nodes \subseteq L \times \{\bar{O}\} \times \mathbb{R}^Z$ is the set of nodes,
- $node_0 = (l_0; \bar{O}_0; u_0)$ is the initial node,
- $\succ = \succ_d \cup \succ_? \cup \succ_{ch} \cup \succ_!$ is the transition relation such that:
  - $(l, \bar{O}, u) \succ_d (l, \bar{O}, u + d)$ if $\forall_{d'}\ 0 \leq d' \leq d \Rightarrow u + d' \in J(l)$
  - $(ch_{ik\_idle}, \bar{O}, u) \succ_! (ch_{ik\_wait\_j}, \bar{O}, u')$: $u \in J(ch_{ik\_idle})$, $u' = [\{cc_{ij}\} \hookrightarrow 0]u$—getting the message from the sending server; transition $(ch_{ik\_idle}, ch_{ij\_wait\_k})$ is labeled with $ch_{ij\_y\_k}?$; asynchronous channel ch_{ij} automaton transition $(ch_{ik\_idle}, \mathbb{R}_{\geq 0}, ch_{ik\_y\_j}?, \{cc_{ij}\}, \{\bar{O}\}, F(\bar{O}) \equiv \bar{O}, ch_{ij\_wait\_k})$
  - $(ch_{ij\_wait\_k}, \bar{O}, u) \succ_{ch} (ch_{ij\_send}, \bar{O}, u')$: $u \in g_{chj}(k) = J(ch_{ij\_wait\_k})$, $u' = [\{cc_{ij}\} \hookrightarrow 0]u$—transition to the location putting the message to the target server (labeled with *delay*); asynchronous channel ch_{ij} automaton transition $(ch_{ij\_wait\_k}, g_{chj}(k), delay, \{cc_{ij}\}, \{\bar{O}\}, F(\bar{O}) \equiv \bar{O}, ch_{ij\_send})$
  - $(ch_{ij\_send}, \bar{O}, u) \succ_? (ch_{ij\_idle}, \bar{O}, u')$: $u \in J(ch_{ij\_send})$, $u' = [\{cc_{ij}\} \hookrightarrow 0]u$—putting the message to the target server; the transition $(ch_{ij\_send}, ch_{ij\_idle})$ is labeled with $ch_{ij\_x}!$; asynchronous channel ch_{ij} automaton transition $(ch_{ij\_send}, \mathbb{R}_{\geq 0}, ch_{ij\_x!}, \{cc_{ij}\}, \{\bar{O}\}, F(\bar{O}) \equiv \bar{O}, ch_{ij\_idle})$.

10.4.5 Translation of CT-IMDS System to Uppaal Timed Automata

The timed system *CT-IMDS* of *N* servers and *K* agents system is defined as (*IMDS*, g_A, g_{ch})—IMDS is the timeless system, time constraints are given as two functions assigning time durations to actions and time delays to asynchronous channels.

$Q = \{ch_{ij\_x},\ ch_{ij\_y\_k} | i = 1..K,\ j = 1..N,\ \mathrm{k} = 1..N\}$ is the set of synchronous channels for implementation of IMDS asynchronous channels. The asynchronous channel ch_{ij_y} is for receiving messages of agent a_i at the server s_j. The channel $ch_{ij\_y\_k}$ is for sending messages of the agent a_i from the server s_k to the sever s_j.

The asynchronous channel between the pair of servers (s_k, s_j) consists of synchronous channels $ch_{ij\_y\_k}$, ch_{ik_x}.

The set of all locations of all servers is $L_S = \cup_{j=1..N} P_{sj}$.

The set of all locations of all asynchronous channels is $L_{CH} = \cup_{i=1..K,\, j=1..N} CH_{ij}$.

$L = L_S \cup L_{CH}$—the set of all locations.

$CS = \{cs_1,\ldots,cs_N\}$ is the set of *hidden clocks* of servers s_i.

$CC = \{cc_{ij} | i = 1..K,\ j = 1..N\}$ is the set of clocks of asynchronous inter-server channels.

$Z = CS\ \cup CC$.

g_Λ: $\Lambda \rightarrow$ *constr* (action time constraint guard) is the function mapping actions to subranges, it is the sum of all functions for individual servers.

The action invariant $J_\Lambda(\lambda)$ is cs_j in the subrange (*true*, 0, x_3, x_4) of $g_\Lambda(\lambda) = (x_1, x_2, x_3, x_4)$, $\lambda = (((a, s_j, r),(s_j, p)),((a, s_k, r'),(s_j, p'))) \in \Lambda$, for all $cs \neq cs_j$ the scope is $\mathbb{R}_{\geq 0}$.

g_{ch}: $S \times S \rightarrow$ *constr* (asynchronous channel delay time constraint guard) is the function mapping all pairs of (*agent*, *target server*) (asynchronous communication channels) to subranges, just like g_Λ.

The asynchronous channel invariant $J_{ch}(ch_{ij})$ in the subrange (*true*, 0, x_3, x_4) of $g_{ch}(ch_{ij}) = (x_1, x_2, x_3, x_4)$, $i = 1..K$, $j = 1..N$, for all $cc \neq cc_{ij}$ the scope is $\mathbb{R}_{\geq 0}$.

$\bar{Y}_p \in P_{s1} \times \cdots \times P_{sN}$, where $P_{si} \subseteq \{p_{sjx}(p_j),\ p_{sj\lambda d}(p_{sj},\lambda,g_\Lambda(\lambda)),\ p_{sj\lambda y}(p_{sj},\lambda) | p_j \in P(s_j),\ j = 1..N,\ \lambda \in \Lambda\}$—the vector of current locations of all server automata. The initial vector of states of servers $\bar{Y}_{p\_ini} = (p_{sjx}(p_j) | p_j \in P(s_j) \cap P_{ini},\ j = 1..N)$.

$\bar{Y}_{ch} \in CH_{11} \times \cdots \times CH_{KN}$, where $CH_{ij} = \{ch_{ij\_idle},\ ch_{ij\_wait\_k},\ ch_{ij\_send} | k = 1..N\} | i = 1..K,\ j = 1..N$—the vector of current locations of asynchronous channels for sending messages to all servers, for every agent. In practice, only used asynchronous channels must be implemented, i.e.:

- For every agent, only the asynchronous channels for sending messages to the servers visited by the agent must be implemented.
- In an asynchronous channel, locations $ch_{ij\_wait\_k}$ must be implemented only for servers s_k sending messages to the server s_j.

$\bar{Y}_{ch\_ini}$ is the initial vector of asynchronous channel locations, $\bar{Y}_{ch\_ini}\ [i, j] \in \{ch_{ij\_idle}, ch_{ij\_send}\}$: if the initial message for the agent a_i is directed to the server s_j, then the initial location of asynchronous channel ch_{ij} is ch_{ij_send}, otherwise ch_{ij_idle}.

$\bar{Y} = (\bar{Y}_p, \bar{Y}_{ch})$—the global location. $\bar{Y}_{ini} = (\bar{Y}_{p\_ini}, \bar{Y}_{ch\_ini})$—the global initial location.

Location invariant functions are composed into a common function over location vectors $J(\bar{Y}) = J(l_1) \wedge J(l_2) \wedge \ldots, l_i \in L, i = 1..card(L)$. For timeless locations $P_{sjx} \cup P_{sjy}$ of the server s_j, the invariant $J(l \in P_{sjx} \cup P_{sjy}) = \mathbb{R}_{\geq 0}$ for the clock cs_j. The invariant for timed locations P_{sjd} $J(l \in P_{sjd}) = J_\Lambda(\lambda)$, $\lambda \in \Lambda$ for the clock cs_j (only for the actions in the server s_j: $\lambda = (((a, s_j, r),(s_j, v)),((a, s_k, r'),(s_j, v')))$. For timeless asynchronous channel locations CH_{ij_idle}, CH_{ij_send}, $J(l \in CH_{ij\_idle} \cup CH_{ij\_send}) = \mathbb{R}_{\geq 0}$ for the clock cc_{ij}. The invariant for timed location $ch_{ij\_wait\_k}$ $J(l) = J_{ch}(k, j)$ for the clock cc_{ij}.

$O = \{o_{ij} | i = 1..K, j = 1..N\}$—the set of variables containing services called by messages. Sending a message (a_i, s_j, r) is implemented as setting the value of the variable o_{ij} and simultaneous executing the *send* operation on corresponding channel $ch_{ik\_y\_j}!$. Accepting the message (execution of the action with this message in input) is simultaneous resetting the variable to the neutral value (outside the set of services R) and executing the receive operation on corresponding channel $ch_{ij\_x}?$.

$W_j = \{w_{1j}, w_{2j}, \ldots\}$, $w_{ij} \in R \cup \{novalue\}$, $i = 1..card(R)$—the set of values implementing the set of services of the server s_j. In practice, only services used in messages directed to s_j can be used.

$\bar{O} = (w_{ij}, \ldots) | i = 1..K, j = 1..N; w_{ij} \in W_i \cup \{novalue\}$—the vector of values of variables implementing messages (w_{ij} is the service of messages of the agent a_i pending at the server s_j); for the agent a_i, at most one of the variables w_{ij}, $j = 1..N$ has a value other than *novalue*: exactly one if the agent runs, all are *novalue* if the agent is terminated. In practice, for any agent only servers visited by the agent are represented.

$\bar{O}_0 = (w_{ij0}, \ldots) | i = 1..K, j = 1..N; w_{ij} \in W_i \cup \{novalue\}$—the initial vector of services of pending messages; for the agent a_i, exactly one of the variables w_{ij}, $j = 1..N$ has value other than *novalue*. For $m_i \in M_{ini}$, $m = (a_i, s_j, r)$, $i = 1..K$, $j \in 1..N$, $w_{ij} = \mathrm{r}$, $\forall_{k \neq j}\, w_{ik} = novalue$.

$\bar{Y}[l_i / l_i']$ denotes the location vector where the ith element l_i of $\bar{Y}$ has been replaced by l_i'.

E_{sj}, $j = 1..N$ is the set of transitions of server automaton s_j. E_{CHij}, $i = 1..K$, $\mathrm{j} = 1..N$ is the set of transitions of asynchronous channel automaton CH_{ij}. Labels λ and *delay* are replaced by the symbol of "internal action" τ.

10.4.6 Semantics of CT-IMDS System Implemented as Uppaal Timed Automata

Let $CT\text{-}IMDS = (IMDS, g_\Lambda, g_{ch})$.

Let u—the valuation of all clocks, u_0—all clocks equal 0.

The semantics is defined as a Labeled Transition

$$\mathsf{LTS} = \langle Node;\ node_0;\ \succ \rangle, \tag{10.7}$$

where:

- $Node \in (P_{s1} \times \cdots \times P_{sN}) \times (CH_{11} \times \cdots \times CH_{KN}) \times (R \cup \{novalue\})^{card(A)*card(S)} \times \mathbb{R}^{CS \cup CC}$ is a *node*,
- $node_0 = (\bar{Y}_0;\ \bar{O}_0;\ u_0)$ is the *initial node*,
- $\succ = \succ_d \cup \succ_? \cup \succ_\lambda \cup \succ_! \cup \succ_{ch}$ is the *transition relation* defined by:

  - $(\bar{Y}, \bar{O}, u) \succ_d (\bar{Y}, \bar{O}, u + d)$ if $\forall_{d'}\ 0 \le d' \le d \Rightarrow u + d' \in J(\bar{Y})$
  - $(\bar{Y}, \bar{O}, u) \succ_? (\bar{Y}[p_{sjx}(p_j)/p_{sj\lambda d}(p_{sj}, \lambda, g_\Lambda(\lambda)), ch_{ij\_send}/ch_{ij\_idle}], \bar{O}, u')$ if there exists $\lambda = ((m, p_j),(m', p'_j)) \in \Lambda$ such that $m = (a_i, s_j, r)$, $m' = (a_i, s_k, r')$, $\bar{O}[i, j] = r$, $u \in J(\bar{Y}[p_{sjx}(p_j)/p_{sj\lambda d}(p_{sj}, \lambda, g_\Lambda(\lambda)), ch_{ij\_send}/ch_{ij\_idle}])$, $u' = [\{cs_j, cc_{ij}\} \hookrightarrow 0]u$—acceptance of the incoming message; note that the transitions $(p_{sjx}(p_j), p_{sj\lambda d}(p_{sj},\lambda,g_\Lambda(\lambda)))$ and $(ch_{ij\_send}, ch_{ij\_idle})$ synchronize od the channel ch_{ij_x} with labels $ch_{ij\_x}?$ and $ch_{ij\_x}!$;
    s_j transition $(p_{sjx}(p_j), \mathbb{R}_{\ge 0}, ch_{ij\_x}?, \{cs_j\}, \{\bar{O}[i, j] = r\}, F(\bar{O}) \equiv \bar{O}, p_{sj\lambda d}(p_{sj}, \lambda,g_\Lambda(\lambda)))$
    ch_{ij} transition $(ch_{ij\_send}, \mathbb{R}_{\ge 0}, ch_{ij\_x}!, \{cc_{ij}\}, \{\bar{O}\}, F(\bar{O}) \equiv \bar{O}, ch_{ij\_idle})$
  - $(\bar{Y}, \bar{O}, u) \succ_\lambda (\bar{Y}[p_{sj\lambda d}(p_{sj}, \lambda, g_\Lambda(\lambda))/p_{sj\lambda y}(p_{sj}, \lambda), \bar{O}', u')$ for $\lambda = ((m, p_j), (m', p'_j)) \in \Lambda$ such that $m = (a_i, s_j, r)$, $m' = (a_i, s_k, r')$, $\bar{O}'[i, j] = novalue$, $\bar{O}'[i, k] = r'$, $u \in g_\Lambda(\lambda) = J(\bar{Y}[p_{sj\lambda d}(p_{sj}, \lambda, g_\Lambda(\lambda))/p_{sj\lambda y}(p_{sj}, \lambda)])$, $u' = [\{cs_j\} \hookrightarrow 0]u$—transition to the location sending the next message (labeled with τ);
    s_j transition $(p_{sj\lambda d}(p_{sj}, \lambda, g_{\Lambda sj}(\lambda)), g_{\Lambda sj}(\lambda), \tau, \{cs_j\}, \{\bar{O}\}, F(\bar{O}) = \bar{O}/[i, k] = r', p_{sj\lambda y}(p_{sj}, \lambda))$
  - $(\bar{Y}, \bar{O}, u) \succ_! (\bar{Y}[p_{sj\lambda y}(p_{sj}, \lambda)/p_{sjx}(p'_j), ch_{ik\_idle}/ch_{ik\_wait\_j}], \bar{O}', u')$ for $\lambda = ((m, p_j),(m', p'_j)) \in \Lambda$ such that $m = (a_i, s_j, r)$, $m' = (a_i, s_k, r')$, $u \in J(\bar{Y}[p_{sj\lambda y}(p_{sj},\lambda)/p_{sjx}(p'_j), ch_{ik\_idle} / ch_{ik\_wait\_j}])$, $u' = [\{cs_j, cc_{ik}\} \hookrightarrow 0]u$—transition to the next state and sending the next message; note that the transitions $(p_{sj\lambda y}(p_{sj}, \lambda), p_{sjx}(p'_j))$ and $(ch_{ik\_idle}, ch_{ik\_wait\_j})$ synchronize on the channel $ch_{ij\_y\_k}$ with labels $ch_{ij\_y\_k}!$ and $ch_{ij\_y\_k}?$;
    s_j transition $(p_{sj\lambda y}(p_{sj}, \lambda), \mathbb{R}_{\ge 0}, ch_{ij\_y\_k}!, \{cs_j\}, \{\bar{O}\}, F(\bar{O}) \equiv \bar{O}, p_{sjx}(p'_j))$
    ch_{ij} transition $(ch_{ik\_idle}, \mathbb{R}_{\ge 0}, ch_{ik\_y\_j}?, \{cc_{ij}\}, \{\bar{O}\}, F(\bar{O}) \equiv \bar{O}, ch_{ij\_wait\_k})$
  - $(\bar{Y}, \bar{O}, u) \succ_{ch} (\bar{Y}[ch_{ij\_wait\_k}/ch_{ij\_send}], \bar{O}, u')$ $u \in g_{ch}(k, j) = J(\bar{Y}[ch_{ij\_wait\_k}/ch_{ij\_send}])$, $u' = [\{cc_{ij}\} \hookrightarrow 0]u$; ch_{ij} transition $(ch_{ij\_wait\_k}, g_{ch}(k, j), \tau, \{cc_{ij}\}, \{\bar{O}\}, F(\bar{O}) \equiv \bar{O}, ch_{ij\_send})$
  - If both $\succ_d$ and ($\succ_?$ or $\succ_!$) are possible from given node $(\bar{Y}, \bar{O}, u)$, then $\succ_d$ from the node is not included into LTS (urgent channels).

10.5 Example of Timed IMDS System—Two Semaphores

To show the operation of the Timed IMDS specification, the "2 semaphores" system is used. The timed version of the system is presented in Fig. 10.9. Several types of time constraints are imposed on the system:

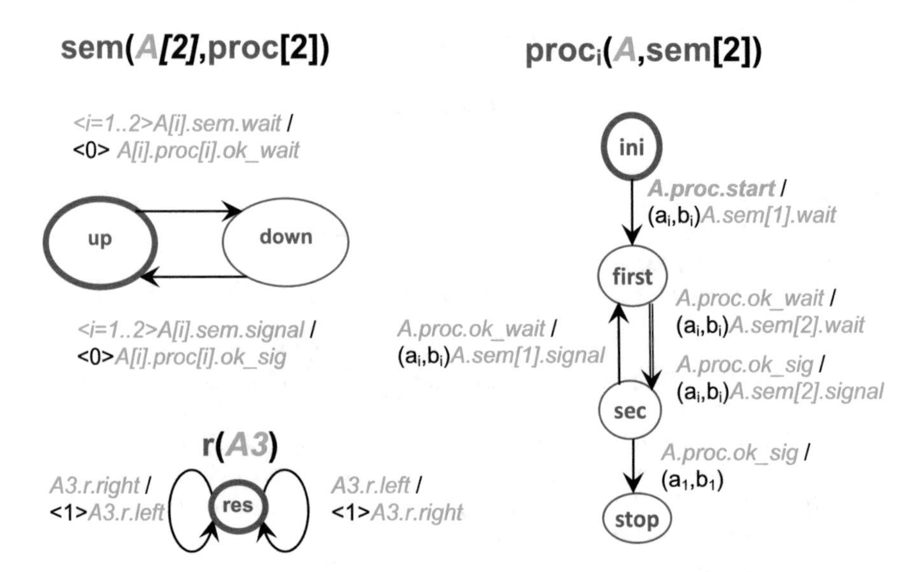

Fig. 10.9 "Two semaphores" example with time constraints: <number> or (number, number) after/—lower and upper limit of action duration

- The actions duration in `proc[1]` is a_1 to b_1 time units, while in `proc[2]` a_2 to b_2 time units (lines 18–25 in the code),
- The actions duration in `sem` servers is instant (0 time units, lines 10–12 in the code below),
- The actions duration in *r* server is 1 time unit (lines 32, 33),
- The time delay of all channels is between 1 and 2 time units.

If the duration of actions in both servers of type *proc* is equal (`a1=a2, b1=b2`, lines 2-5), the system operates just like the timeless system described in Sect. 5.1 (a deadlock occurs). But if one of the servers acts significantly faster than the other one (`a2=7, b2=8`), the deadlock disappears. This happens because the first server catches two *wait* operations before the second one issues the first `wait`.

The source code of the system with time constraints is given below:

```
 1. system two_semaphores_t

 2. #DEFINE a1=0
 3. #DEFINE a2=0
 4. #DEFINE b1=1
 5. #DEFINE b2=1

 6. server:   sem(agents A[2];servers proc[2]),
 7. services {wait, signal},
 8. states   {up, down},
 9. actions  {
10.    <j=1..2>{A[j].sem.wait, sem.up}   -> <0> {A[j].proc[j].ok_wait, sem.down},
11.    <j=1..2>{A[j].sem.signal, sem.down} -> <0> {A[j].proc[j].ok_sig, sem.up},
12.    <j=1..2>{A[j].sem.signal, sem.up}-> <0> {A[j].proc[j].ok_sig, sem.up},
13. };

14. server:  proc(agents A;servers sem[2]),
15. services {start, ok_wait, ok_sig},
16. states          {ini, first, sec, stop},
17. actions  {
18.    ?1 {A.proc.start, proc.ini}       ->(a1,b1) {A.sem[1].wait, proc.first},
19.    ?2 {A.proc.start, proc.ini}       ->(a2,b2) {A.sem[1].wait, proc.first},
20.    ?1 {A.proc.ok_wait, proc.first}   ->(a1,b1) {A.sem[2].wait, proc.sec},
21.    ?2 {A.proc.ok_wait, proc.first}   ->(a2,b2) {A.sem[2].wait, proc.sec},
22.    ?1 {A.proc.ok_wait, proc.sec}     ->(a1,b1) {A.sem[1].signal, proc.first},
23.    ?2 {A.proc.ok_wait, proc.sec}     ->(a2,b2) {A.sem[1].signal, proc.first},
24.    ?1 {A.proc.ok_sig, proc.first}    ->(a1,b1) {A.sem[2].signal, proc.sec},
25.    ?2 {A.proc.ok_sig, proc.first}    ->(a2,b2) {A.sem[2].signal, proc.sec},
26.       {A.proc.ok_sig, proc.sec}      ->(a1,b1) {proc.stop},
27. };

28. server:  r(agents A3),
29. services {left, right},
30. states   {res},
31. actions  {
32.    {A3.r.left, r.res}  -> <1> {A3.r.right, r.res},
33.    {A3.r.right, r.res} -> <1> {A3.r.left, r.res},
34. };

35. servers sem[2],proc[2],r;
36. agents A[2],A3;

37. channels {(1,2)};

38. init      ->     {
39.           proc[1](A[1],sem[1,2]).ini,
40.           proc[2](A[2],sem[2],sem[1]).ini,
41.           <j=1..2>sem[j](A[1..2],proc[1..2]).up,
42.           r(A3).res,

43.           A[1].proc[1].start,
44.           A[2].proc[2].start,
45.           A3.r.left,
46. }.
```

If a_1=a_2, b_1=b_2, but we change the time delay of individual channels, namely between *proc2* and the rest of the system:

```
channels {          (1,2),
                    proc[2]->sem(9,10),
                    sem->proc[2](9,10)};
```

then the deadlock disappears.

10.6 Example of Timed IMDS System—Automatic Vehicle Guidance System

The example of "intersection" (Sect. 5.3), both in version 3-way and 4-way, can be successfully verified under Uppaal. However, the example equipped with time constraints causes memory overrun (8 GB). The simpler example of the automatic vehicle guidance system (Sect. 5.4) gives this possibility.

Time constraints applied to CT-IMDS variant are:

- The time of occupation of the road markers and parking lots (the time duration of the actions changing states `occupied` to `free`) is between x and y.
- The time of traveling between the markers and lots (the time delay of every channel) is between x and y.

The "almost no constraints" model, similar to a timeless system, is achieved by setting $x = 1$ and $y = 10$. This gives a deadlock like in the timeless system described in Sect. 5.4.

The simultaneous movement of two AMPs, starting almost synchronously, is modeled by the constraints $x = 1$ and $y = 2$. In this situation, the deadlock disappears—the AMPs evade in *`markerM`* using *`lotM`*, but no evading necessity appears in *`markerE[1]`* and *`markerE[2]`* (time constraints do not allow the AMPs to meet there).

10.7 Verification of the Production Cell

An attempt to verify the timed version of the Production Cell (Sect. 5.6) failed due to memory overrun. In the literature, some other real time verification approaches, based on synchronous modeling, succeeded (Benghazi Akhlaki et al. 2007; Behrmann et al. 2006). Our fail comes from explicit modeling of asynchronous channels built from two synchronous channels, for every pair (*agent*, *target server*). This significantly increases the size of the reachability space for model checking.

The future implementation of our own timed verifier will be free from such disadvantages, because asynchrony is modeled directly in the succession of actions without any additional constructs.

In addition, we are currently working on a timed version of the non-exhaustive search in the reachability space in the "2 vagabonds" algorithm (Chap. 11).

References

Alur, R., & Dill, D. L. (1994). A theory of timed automata. *Theoretical Computer Science, 126*(2), 183–235. https://doi.org/10.1016/0304-3975(94)90010-8.

Behrmann, G., Cougnard, A., David, A., Fleury, E., Gulstrand Larsen, K., & Lime, D. (2006). Uppaal-Tiga: Timed games for everyone. In *Nordic Workshop on Programming Theory (NWPT'06), Reykjavik, Iceland*, October 18–20, 2006. URL: https://hal.archives-ouvertes.fr/hal-00350470/document.

Behrmann, G., David, A., & Larsen, K. G. (2006). *A tutorial on Uppaal 4.0*. Aalborg, Denmark. URL: http://www.it.uu.se/research/group/darts/papers/texts/new-tutorial.pdf.

Benghazi Akhlaki, K., Capel Tuñón, M. I., Holgado Terriza, J. A., & Mendoza Morales L E. (2007). A methodological approach to the formal specification of real-time systems by transformation of UML-RT design models. *Science of Computer Programming, 65*(1), 41–56. https://doi.org/10.1016/j.scico.2006.08.005.

Bérard, B. (2013). An introduction to timed automata. In *Control of discrete-event systems* (pp. 169–187). London, UK: Springer. https://doi.org/10.1007/978-1-4471-4276-8_9.

Bérard, B., Cassez, F., Haddad, S., Lime, D., & Roux, O. H. (2005). Comparison of the expressiveness of timed automata and time Petri nets. In *Third International Conference FORMATS 2005, Uppsala, Sweden* (pp. 211–225), September 26–28, 2005. Berlin: Springer. https://doi.org/10.1007/11603009_17.

Bowman, H. (2001). Time and action lock freedom properties for timed automata. In *21st International Conference on Formal Techniques for Networked and Distributed Systems, FORTE 2001, Cheju Island, Korea* (pp. 119–134), August 28–31, 2001. Boston: Kluwer Academic Publishers. https://doi.org/10.1007/0-306-47003-9_8.

Bowman, H., Gomez, R., & Su, L. (2005). A tool for the syntactic detection of zeno-timelocks in timed automata. *Electronic Notes in Theoretical Computer Science, 139*(1), 25–47. https://doi.org/10.1016/j.entcs.2005.09.006.

Cassez, F., & Roux, O. H. (2006). Structural translation from time Petri nets to timed automata. *Journal of Systems and Software, 79*(10), 1456–1468. https://doi.org/10.1016/j.jss.2005.12.021.

Emerson, E. A., Mok, A. K., Sistla, P., & Srinivasan, J. (1992). Quantitative temporal reasoning. *Real-Time Systems, 4*(4), 331–352. https://doi.org/10.1007/bf00355298.

Frossl, J., Gerlach, J, & Kropf, T. (1996). An efficient algorithm for real-time symbolic model checking. In *Proceedings ED&TC European Design and Test Conference, Paris, France* (pp. 15–20), March 11–14, 1996. New York, NY: IEEE Computer Society Press. https://doi.org/10.1109/edtc.1996.494120.

Gluchowski, P. (2009). Languages of CTL and RTCTL Calculi in real-time analysis of a system described by a fault tree with time dependencies. In *2009 Fourth International Conference on Dependability of Computer Systems, DepCoS–RELCOMEX '09, Brunów, Poland* (pp. 33–41), June 30–July 2, 2009. New York, NY: IEEE. https://doi.org/10.1109/depcos-relcomex.2009.12.

Gómez, R., & Bowman, H. (2007). Efficient detection of zeno runs in timed automata. In *5th International Conference on Formal Modeling and Analysis of Timed Systems (FORMATS 2007), Salzburg, Austria* (pp. 195–210), October 3–5, 2007, LNCS 4763. Berlin, Heidelberg: Springer. https://doi.org/10.1007/978-3-540-75454-1_15.

Hadjidj, R., Boucheneb, H., & Hadjidj, D. (2007). Zenoness detection and timed model checking for real time systems. In K. Barkaoui & M. Ioualalen (Eds.), *First International Conference on Verification and Evaluation of Computer and Communication Systems (VECoS'07), Algiers, Algeria* (pp. 120–134), May 5–6, 2007. Swinton, UK: British Computer Society. url: https://www.bcs.org/upload/pdf/ewic_ve07_s3paper3.pdf.

Holzmann, G. J. (1995). Tutorial: Proving properties of concurrent systems with SPIN. In *6th International Conference on Concurrency Theory (CONCUR'95, Philadelphia, PA)* (pp. 453–455), August 21–24, 1995. Berlin, Heidelberg: Springer. https://doi.org/10.1007/3-540-60218-6_34.

Krystosik, A., & Turlej, D. (2006). EMLAN: A language for model checking of embedded systems software. In *IFAC Workshop on Programmable Devices and Embedded Systems, Brno, Czech Republic* (pp. 126–131), February 14–16, 2006. Amsterdam: Elsevier Science. https://doi.org/10.1016/S1474-6670(17)30171-4

Lime, D., & Roux, O. H. (2006). Model checking of time petri nets using the state class timed automaton. *Discrete Event Dynamic Systems, 16*(2), 179–205. https://doi.org/10.1007/s10626-006-8133-9.

Lindahl, M., Pettersson, P., & Yi, W. (2001). Formal design and analysis of a gear controller. *International Journal on Software Tools for Technology Transfer, 3*, 353–368. https://doi.org/10.1007/s100090100048.

Popescu, C., & Martinez Lastra, J. L. (2010). Formal methods in factory automation. In J. Silvestre-Blanes (Ed.), *Factory automation* (pp. 463–475). Rijeka, Croatia: InTech. https://doi.org/10.5772/9526.

Ruf, J., & Kropf, T. (2003). Symbolic verification and analysis of discrete timed systems. *Formal Methods in System Design, 23*(1), 67–108. https://doi.org/10.1023/A:1024437214071.

Winskel, G., & Nielsen, M. (1995). Models for concurrency. In S. Abramsky, D. M. Gabbay, & T. S. E. Maibaum (Eds.), *Handbook of logic in computer science* (Vol. 4, pp. 1–148). Oxford, UK: Oxford University Press. ISBN: 0-19-853780-8

Chapter 11
2-Vagabonds: Non-exhaustive Verification Algorithm

Static deadlock detection methods suffer from space explosion problem. Model checking and other static analysis techniques are very effective in verification, many of them have a linear complexity to the size of the reachability space. However, the elaboration of the space is time-consuming (usually exponential) and takes a large amount of memory.

Some methods fight space explosion: symbolic model checking, abstraction etc. There are also methods that give an approximate solution, based on non-exhaustive heuristics for finding deadlocks. Some of them are discussed below. Finding a deadlock proves its existence, the lack of finding a does not guarantee its absence. Nowadays, we are used to such techniques in computer science, for example machine learning give both false positive and false negative solutions (Batista et al. 2004).

There are several non-exhaustive verification methods. Three of them are described in multiple papers: Ant Colony Optimization—ACO (Francesca et al. 2011), Genetic Algorithms—GA (Godefroid and Khurshid 2002) and A* search (Francesca et al. 2011) (they are mixed in some solutions). The methods consist in construction of many reachable paths, hoping that at least one of them reaches a deadlock. Non-exhaustive methods were developed originally for other purposes, therefore some elements of them are useless. For example, drawing the chromosome genes (system transitions play this role) does not guarantee the construction of a reachable path. In addition, those methods use sets of parameters which values are not obvious. In GA, they are: population size, chromosome length, gene crossing manner, mutation rate, fitness function, etc. In ACO there are number of ants, pheromone evaporation rate, parameters for pheromone deposit etc. It is not easy to define the parameter values for a given case.

All algorithms based on the construction of paths have a general drawback: similar beginning of the paths. The reachability space is often extremely branched, but its initial part consists of only several subpaths. Therefore, the initial subpaths

© Springer Nature Switzerland AG 2020

W. B. Daszczuk, *Integrated Model of Distributed Systems*, Studies in Computational Intelligence 817, https://doi.org/10.1007/978-3-030-12835-7_11

occur in multiple generated paths. In addition, many systems contain a starting part, in which the initial activities are performed (we call it *leader* (Daszczuk 2018a)). A deadlock-free leader occurs in all paths without purpose (from the point of view of deadlock detection).

The methods described above do not distinguish between deadlocks and termination, as described in Chap. 2. For example successful termination of solitaire game is found just as if it were a deadlock (Gradara et al. 2006).

The advantages of the mentioned algorithms are:

- Paths are constructed dynamically as the algorithm runs, so the global reachability space need not be elaborated.
- The successful path (leading to the deadlock) is a ready-to-use counterexample, while in the case of exhaustive search, the path from the initial node to the deadlock must be found.

Finding deadlocks using those methods is very effective, but we think it should be extended. Distributed IoT systems contain reactive servers that offer a variety of services and operate in infinite loops, and device controllers perform certain coordinated behavior, then finish their work. In such systems, the deadlock may not be total. In addition, total distributed termination is not expected because of looping servers. Instead, a partial deadlock and termination are very much needed to be verified. However, concurrency degree in such systems is even greater than in centralized systems, because some services are often spread over distributed servers.

The mentioned papers on heuristic methods show deadlock detection in several benchmarks (most often Dijkstra philosophers (Dijkstra 1959)). The drawback of the papers is that they exhibit detection when the result is known in advance. However, in practice, the user is interested in verifying the system in which the result is unknown. That is the sense of verification. The papers do not describe such procedures, especially the time of verification of deadlock-free system is not given.

That's why we needed an efficient algorithm for partial deadlock and partial termination verification, using a non-exhaustive search. We developed our own algorithm which has the following features:

- Location of partial deadlocks that are not just nodes without outgoing transitions.
- Distinguishing deadlocks from distributed termination.
- Having as few parameters as possible, preferably zero.
- Generating of leader subpaths less often than for all generated paths.

This former two features are achieved thanks to the definition of partial deadlocks in the IMDS, which consists in general temporal formulas, unrelated to the structure of verified system. The formulas for distributed termination differ from the deadlock formulas. The latter two features are achieved as specific properties of the new algorithm, which is based on wandering through reachability space, constructed on-the-fly. Since each search consists of two phases, which can run almost independently, we call the algorithm *2-vagabonds*.

The new algorithm searches the reachability space in two phases. The first phase searches for nodes in which a subset of processes are disabled, the second phase checks if these processes can be enabled in the future. Because both phases use non-exhaustive search, the result is unsure and both false positives and false negatives are possible. This is typical for non-exhaustive search, for example in Godefroid and Khurshid (2002) the authors agree to lose the proper result in up to 70% cases.

11.1 Related Work

State space explosion is the main obstacle in model checking. There are many methods to decrease this feature, such as symbolic model checking (Clarke et al. 1996), bounded model checking (Jard and Jéron 1992), compositional model checking (Clarke et al. 1989; Santone 2002), local model checking (Stirling and Walker 1991), partial order reduction (Penczek et al. 1999), abstraction (Clarke et al. 1994) and slicing (Hatcliff et al. 2000).

A non-exhaustive search fights the explosion problem at the expense of uncertainty of the result. There are three main methods: Ant Colony Optimization (ACO), A* graph search, and Genetic Algorithms (GA).

ACO solutions are rather optimization algorithms than search ones (Francesca et al. 2011; Chicano and Alba 2008). They work like an ant, which if finds a target, i.e. a deadlock or termination, puts a pheromone that directs other ants. Therefore, ACO is ideal for finding shortest counterexample after finding a deadlock. However, during the verification, it is difficult to determine which path more likely leads to a deadlock than another path. If the ACO algorithm is used for search rather than for optimization, some paths must seem more attractive to ants than other paths.

In Alba and Chicano (2007), the heuristic is used that prefers a large number of enabled transitions in each node (configuration in our IMDS formalism). It is not explained why this should lead to a deadlock faster. Indeed, in that paper three benchmarks out of five give worse results for the described heuristic version than the basic one (in memory usage and in verification time).

A* uses the heuristic evaluation function, associated with each node or transition (Francesca et al. 2011; Gradara et al. 2006). The applied heuristic must be admissible, i.e., it must underestimate the cost. The algorithm prefers paths that seem better, following nodes or transitions with a lower value of the evaluation function, which should accelerate the achievement of the goal. It is difficult to define a function that prefers paths leading to an unknown deadlock. In Gradara et al. (2006), such a heuristic is proposed for specification in CCS (Milner 1984), with the preference for deadlocks of communication operations. The heuristic is addressed to total deadlocks. The formalism of CCS is synchronous, which in our opinion is inadequate for modeling distributed systems.

Gradara et al. (2006) proposed a Greedy algorithm, similar to A*, using a non-admissible heuristic. It tends to find quite short (but not optimal) counterexamples faster than A*.

Genetic Algorithm generates paths of execution as chromosomes (Godefroid and Khurshid 2002; Alba and Troya 1996; Yousefian et al. 2014). Genes are choices of transitions between nodes (configurations in IMDS). The above problems with evaluating transitions in ACO and A* are similar to the troubles with the fitness function of the chromosomes in GA. If we investigate for a deadlock, finding it simply finishes the algorithm and no further path evaluation is needed. If we search for a counterexample having given features (for example length), a more subtle evaluation is needed. For example, reinforcement learning applied to GA is described in Pira et al. (2017).

11.2 2-Vagabonds Algorithm

11.2.1 Total Deadlock

We begin the description of our new algorithm with total deadlock detection, which requires only one vagabond. The rules are as follows:

- A path from the initial configuration is constructed. The path consists of configurations connected by actions. The path is constructed from actions one after the other, with a random selection of the action among the actions enabled in the configuration. In the configuration T, the action $(m,p)\Lambda(m',p')$ is enabled if $\{m,p\}\subset T$.
- If the deadlocked configuration is reached, the algorithm stops.
- If a configuration already in the path is found, it means a loop. In this situation, the path is truncated by a random number of configurations. From the last configuration of the truncated path, the action is selected randomly, excluding the recently used action. If only one action is enabled in the configuration, the path is truncated again (by one configuration).

Note that the algorithm is extremely simple and does not contain any parameters. However, to successfully run it, one parameter is needed: how long must we construct a path if no deadlock is found. Our first attempt was to limit the number of algorithm steps (number of path extensions) to N:

for server deadlocks:

$$N = card(\Lambda) * \left\lceil \log_2\left(\prod_{k=1}^{n} card(\Lambda s_k)\right)\right\rceil * \lceil \log_2(n)\rceil * \lceil \log_2(m)\rceil, \qquad (11.1)$$

for agent deadlocks and termination:

$$N = card(\Lambda) * \left\lceil \log_2\left(\prod_{k=1}^{m} card(\Lambda a_k)\right)\right\rceil * \lceil \log_2(n)\rceil * \lceil \log_2(m)\rceil,$$

where Λs_k is the set of actions of the server process s_k, Λa_k is the set of actions of the agent process a_k, n_S and n_A are the number of servers and the number on agents, respectively. The formulas say that we take the number of actions multiplied by the logarithm of the product of the numbers of actions in individual server (or agent) processes, and multiplied by logarithms of the number of servers and the number of agents. The factor after *card*(Λ) comes from the assumption, that overall server states communicate in pairs, but not every pair. The logarithm comes from the assumption that the more states, the smaller part of their pairs communicate. The last two factors result from the assumption that servers communicate in pairs and agents communicate in pairs.

However, in some cases we found a few false negatives, i.e., existent deadlock not found. We introduced the parameter D, which is the factor by witch the value N is multiplied. We suggest $D = 4$, because it is sufficient for most searches concerning single-process deadlock or termination. Searching for a total deadlock requires a larger D in some cases.

In the total server (communication) deadlock, all server processes take part. In the total agent deadlock, all non-terminated agent processes participate.

```
Vagabond (set of all nodes S): // or all agents A
insert initial configuration T0 do a path Φ;
loop
{
    take a configuration T finishing a path Φ;
    if (no s∈S enabled in T) halt(total deadlock); // or all a∈A
    if (exist actions ΛT prepared in T)
    {
        draw an action λ∈ΛT;
        if (λ leads to a configuration T∈Φ) // a cycle detected
        {
            cut a random number of configurations from path Φ;
            while (last configuration Tlast≠T0 in path Φ has one prepared action)
              cut Tlast from path Φ;
            exclude an action drawn previously in new Tlast from next draw;
        }
        else add T to the path Φ;
    }
    else
    {
        cut a random number of configurations from path Φ;
        while (last configuration Tlast≠T0 in path Φ has one prepared action)
            cut Tlast from path Φ;
        exclude an action drawn previously in new Tlast from next draw;
    }
}
until (D*N steps completed);
halt(no total deadlock);
```

11.2.2 Partial Server Deadlock (Communication Deadlock)

Recall the server deadlock definition for a single server (Sect. 3.7): no action is prepared in the server while there are pending messages at the server. If a partial deadlock for a subset of server processes is searched, the situation must concern all servers in the subset. We use two vagabonds to detect server deadlock. For a given set of servers, first vagabond searches for configurations in which no action is prepared for any server in the subset, and each server in the subset has a pending message. Therefore, 1st vagabond starts the 2nd vagabond from this configuration to check if any of the servers will be enabled in the future. Starting from this configuration, the second vagabond searches for a configuration in which in a server belonging to the set has an enabled action. If no such configuration is found, a partial server deadlock occurs.

In the algorithm, parts specific to server deadlock detection are marked by a box (■).

```
1st_vagabond (set of selected nodes Sx):
insert initial configuration T0 do a path Φ1;
loop
{
    take a configuration T finishing a path Φ1;
■   if (no s∈S enabled in T) halt(total deadlock); // implies partial deadlock
■   if (no s∈Sx is enabled but messages to all s∈Sx are in T)
        start 2nd vagabond from T;
    if (exist actions ΛT prepared in T)
    {
        draw an action λ∈ΛT;
        if (λ leads to a configuration T∈Φ1) // a cycle detected
        {
            cut a random number of configurations from path Φ1;
            while (last configuration Tlast≠T0 in path Φ1 has one prepared action)
                cut Tlast from path Φ1;
            exclude an action drawn previously in new Tlast from next draw;
        }
        else add T to the path Φ1;
    }
    else
    {
        cut a random number of configurations from path Φ1;
        while (last configuration Tlast≠T0 in path Φ1 has one prepared action)
            Cut Tlast from path Φ1;
        exclude an action drawn previously in new Tlast from next draw;
    }
}
until (N steps completed);
wait until (2nd vagabond works);
if (2nd vagabond found a configuration in which an s∈Sx is enabled)
    halt(no partial deadlock);
else halt(partial deadlock);

2nd vagabond:
Tstart is a starting configuration for 2nd vagabond
insert starting configuration Tstart do a path Φ2;
loop
{
    take a configuration T finishing a path Φ2;
■   if (all s∈S not enabled)
■       return(partial deadlock found);
■   if (exist s∈Sx enabled)
■       return(no partial deadlock);
    if (exist actions ΛT prepared in T)
    {
        draw an action λ∈ΛT;
        if (λ leads to a configuration T∈Φ2) // a cycle detected - backtracking
        {
            cut a random number of configurations from path Φ2;
            while (last configuration Tlast≠T0 in path Φ2 has one prepared action)
                cut Tlast from path Φ2;
            exclude an action drawn previously in new Tlast from next draw;
        }
        else add T to the path Φ2;
    }
    else return(partial deadlock found);
}
until (D*N steps completed);
```

11.2.3 Partial Agent Deadlock (Resource Deadlock)

An agent is deadlocked if its current message is pending, but it will never cause an action. Two vagabonds are used again. For a given set of agents, the first vagabond searches for configurations in which no action is enabled for any agent in the set, and no agent in the set is terminated. Starting from this configuration, the second vagabond searches for a configuration in which an action is enabled in an agent belonging to the set. If no such configuration is found, an agent deadlock occurs.

The differences between server deadlock and agent deadlock detection are as follows:

```
1st vagabond (set of selected agents Ax):
…
■     if (no a∈A enabled) halt(total deadlock); // implies partial deadlock
■     if (no a∈Ax is enabled nor terminated in T)
…
if (2nd vagabond found a configuration in which an a∈Ax is enabled)
…

2nd vagabond:
…
■     if (all a∈A not enabled)
■           return(partial deadlock found);
■     if (exist a∈Ax enabled)
■           return(no partial deadlock);
…
```

11.2.4 Partial Distributed Termination

Distributed termination of a subset of agents is successful if all agents in the subset reached their terminating actions. The vagabonds search for:

- Total deadlock—in this case the termination is not achieved.
- A partial deadlock with at least one agent belonging to the subset. The partial deadlock is identified in the same way as in partial agent deadlock detection above, using the same second vagabond as in the case of a partial agent deadlock. If such a partial deadlock is found, the termination is refuted.
- A state in which agents in the subset are all terminated.

In the two former cases, the termination is unsuccessful. In the latter case, the termination is achieved but it does not guarantee that it is inevitable. Therefore, after finding the configuration in which the agents belonging to the subset terminate, a new search begins. After reaching of $D * N$ steps of verification, the algorithm is continued to complete the last search. Four situations can occur:

- Total deadlock (as before)—no termination.

- Partial deadlock (as before)—no termination.
- Termination of all agents in the set—this finally confirms distributed termination.
- None of the above occurred in $D * N$ steps plus final search—no termination.

```
1st_vagabond (set of selected agents Ax):
insert initial configuration T0 do a path Φ1;
loop
{
    take a configuration T finishing a path Φ1;
    if (no a∈A is enabled but there exists a∈Ax not terminated)
        return(partial deadlock found);
    if (all a∈Ax terminated in T)
        if (current step > D*N) return (partial termination);
        else { shrink Φ1 to T0, begin new search; }
    else if (no agent in Ax is enabled)
        start 2nd vagabond from T;
    if (exist actions ΛT prepared in T)
    {
        draw an action λ∈ΛT;
        if (λ leads to a configuration T∈Φ1) // a cycle detected - backtracking
        {
            cut a random number of configurations from path Φ1;
            while (last configuration Tlast≠T0 in path Φ1 has 1 prepared action)
                cut Tlast from path Φ1;
            exclude an action drawn previously in new Tlast from next draw;
        }
        else add T to the path Φ1;
    }
    else
    {
        cut a random number of configurations from path Φ1;
        while (last configuration Tlast≠T0 in path Φ1 has one prepared action)
            cut Tlast from path Φ1;
        exclude an action drawn previously in new Tlast from next draw;
    }
}
until (D*N steps completed from last termination found);
wait until (2nd vagabond works);
if (2nd vagabond found a configuration in which an a∈Ax is enabled)
    halt(no partial termination);
else halt(partial termination);
```

11.2.5 Heuristics

We tried to speed up the search in our algorithm. We found out that having some actions disabled in a process, it can lead to a deadlock in effect. Therefore, we increase the probability of selecting transitions leading to configurations with some nodes/agents disabled. We assign the weight to each action outgoing from a configuration:

- 0—if the backtracking is executed because of a loop: 0 is used for the recently used action,
- 1—otherwise, but:
- *d*—if there are *d* disabled processes in the target configuration of the action,
- *e*—if there are *e* disabled processes in the target configuration of the action, for the processes belonging to the subset for which a partial deadlock is investigated.

All weights are summed for a given action ($1 + d + e$). After assigning the weights, the sum for all actions is normalized to 1, and all action weights are normalized accordingly, which gives the probability of taking individual actions.

Indeed, heuristics are accelerating the finding of deadlock in some cases. Even more important is the fact that there are cases (for example 15 philosophers, see below) in which the deadlock cannot be found in $4*N$ steps without heuristics.

11.2.6 Verification

Our algorithm finds partial deadlocks, but it is not known in advance which processes fall into deadlocks and which do not. Our typical procedure is as follows:

- Check the total deadlock: the partial deadlock into which all processes fall.
- If the first verification does not show deadlock, check whether the individual processes fall info a deadlock. The result of such verification in the Dedan program, in which not all processes fall, is presented in Fig. 11.1a. In this verification, the "together" checkbox in the upper right corner is turned off.
- If there are processes that fall into the deadlock individually, switch the checkbox "together" on and select only deadlock-prone processes for verification (checkboxes on the left), it is shown in Fig. 11.1b. If the processes do not fall into a common deadlock, the result "Deadlock" changes to the result "Ok". In this example, the result "Deadlock" is preserved, which means that it is a common partial deadlock of processes *T*, *R* and *C*.

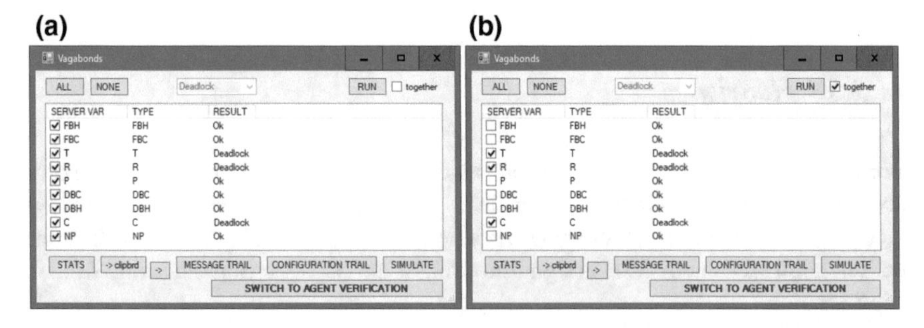

Fig. 11.1 **a** Finding processes falling into a deadlock individually; **b** verification if they fall into a common deadlock

11.3 Experimental Results

11.3.1 The Benchmarks

The algorithm was tested on a set of benchmarks. The first group contains models which can be verified using an exhaustive search:

- "1"—a simple system in which each agent can fall into a deadlock due to lack of resources. It is an interesting example, because in this system the agent falls into the resource deadlock, but no server falls into the communication error. What's more, each agent can fall into a deadlock separately, but a set of more than one process never falls into a common deadlock. 4 servers, 2 agents.
- "2sem"—2 semaphores, (Sect. 5.1) two agents from their own servers invoke P and V on two semaphores, if they call agreeably—no deadlock occurs ("2sem-agr"). Otherwise they call P crosswise—the deadlock is obvious ("2sem-cross"). 5 servers, 3 agents. The deadlock is partial, because the third process doing something else is added.
- "PhX"—Dijkstra philosophers (Dijkstra 1959) (Sect. 5.2), X is their cardinality; every philosopher agent resides on its own server-chair and calls two fork servers: left and right. Chairs and forks are servers ($2X$), philosophers are agents (X). Server deadlock is partial (chair servers are not deadlocked), while agent deadlock is total.
- In addition, three deadlock-free solutions from Sect. 5.2 are verified: asymmetrical—philosophers take their forks in undefined order, and two of them take in specified order: one take his left fork first, while the other one takes his right fork first ("PhX-asym"—$2X$ servers, X agents); with butlers ("PhX-butlers"—$2X + 2$ servers, X agents); taking 2 forks or none ("PhX-2orNo"—$2X + 1$ servers, X agents). Butlers and locker for taking 2 forks at once are additional servers.
- "PCX"—Karlsruhe Production Cell (Daszczuk 2019a) (Sect. 5.6), a set of devices supplying blanks (agents) to be forged in the press, and then taking them back, X is the number of blanks. Blanks are not forged in fact, so they are recycled back in the loop. With proper control no deadlock occurs. Every device in the cell is a server (11 in total), and every blank is an agent. In addition, some technical agents are used to return the devices to the rest position after transporting a blank. There are $X + 6$ agents in total.
- If the above system is tested in an open loop ("PCXopen"), in which the blanks are taken from the container are they are dropped out after forging, a deadlock occurs because the robot exchanges the forged blank with the next one, and the last blank has no pair to exchange with. The press falls into the server deadlock, and, consequently, all devices equipped with technical agents fall into the deadlock. The deadlock is partial because other devices simply remain idle. Each blank falls into the agent deadlock (because the order in which the blanks are entered into a system is arbitrary), but separately. Never two or more blanks

fall into the common deadlock, because only one blank is stuck in the press. Therefore, the agent deadlock is total if $X = 1$, and partial if $X > 1$.
- "Intersection"—Sect. 5.3: 4 crossing quarters and 4 commuter roads are modeled. Vehicles (agents) can take any direction: left, right, straight or turn back. The full model exceeds available memory, which is why the simplified models are verified: with all vehicles driving straight ("Cross-straight"), falling into a deadlock, and all vehicles driving right ("Cross-right"), without deadlock. Every quarter and every commuter road is a server, the vehicles are modeled as agents. Thus, the server deadlock in the former model is partial (commuter roads do not fall into a deadlock), and the agent deadlock is total (all four vehicles fall).
- "AVGS"—Automated Vehicle Guidance System (Czejdo et al. 2016) (Sect. 5.4), the vehicles (agents) are led by road segment controllers (servers). The aim of each vehicle is to achieve the opposite utmost position, but common two-way segments may lead to a deadlock. The basic solution for 6 servers and 2 agents is deadlock-prone. As with "Cross-straight", the server deadlock is partial (only two road segments holding the vehicles are involved, others are idle) and the agent deadlock is total.
- Some other deadlock-free structures of AVGS are used: backing up ("AVGS-goBack"—6 servers, 2 agents), using the overtaking shunt ("AVGS-turnout"—4 servers, 2 agents) and changing the vehicles' decision using alternative paths ("AVGS-twoWays"—4 servers, 2 agents).
- "buf"—bounded buffer (Sect. 9.1) with X elements, Y users ("buf-XxY" (Daszczuk 2019b))—users are both producers and consumers, switching between the roles randomly. If everyone chooses to get from an empty buffer, or if everyone chooses to put to a full buffer, a deadlock occurs.

The second group of benchmarks contains models that cannot be exhaustively verified (the reachability space exceeds the available memory), but the result is known in advance, because they are extensions of problems in first group: "PhX"—for $X = 6$ and more philosophers, "PCX"—for $X = 4$ and more blanks, "PCXopen"—for $X = 7$ or more blanks, "Cross-all" for road crossing with vehicles taking arbitrary direction: left, right, straight or turn back.

The third group of examples are student solutions of synchronization exercises. The result is unknown in advance in these cases. Some of them cannot be verified exhaustively.

11.3.2 Results

The results of deadlock detection are collected in Table 11.1. The deadlock is investigated for every process separately, so number of steps visited by the vagabonds and verification time are given in the average for one server/agent (except for the total deadlock detection). The column "type" specifies the type of deadlock—communication deadlock of *servers* or resource deadlock of *agents*. The letter "t" means finding a total deadlock. All numbers are rounded to 1 digit after point, and all times are rounded to the full seconds.

In some examples, exhaustive verification is compared with 2-vagabonds.

Table 11.2 contains the results of verifying deadlock-free processes. Some benchmarks are included in both Tables 11.1 and 11.2, because in these cases we have processes falling into a deadlock and deadlock-free ones in the same model. The column containing the number of steps for the 1st vagabond is suppressed, because in deadlock-free search this number of steps is exactly maximum number of steps ($D * N$).

Table 11.3 contains the results of the termination checking. Both positive and negative examples are included. Negative result outcomes from falling a process in the set into a deadlock or infinite looping without termination.

Table 11.4 shows the results of student exercise verification. In such cases, the result is not known in advance. "Servers dd" denotes node deadlock search while "Agents dd" denotes agent deadlock search.

One of the main assumptions in design of 2-vagabonds algorithm is avoiding repeatedly investigating of leader subpaths. Figure 11.2 illustrates this. The number of steps in the algorithm runs is divided into segments of length being consecutive powers of two, minus numbers of steps in previous segments. This gives "nominal" numbers of steps in segments 1, 1, 2, 4, 8, 16, 32, … . The numbers of steps in other algorithms give the same powers of 2, multiplied by the number ants in ACO/ individuals in GA etc. For example, the ACO algorithm running 10 ants forth and back 100 times has the following numbers of steps in the segments: 1000, 1000, 2000, 4000, 8000, etc. Genetic algorithm with 100 individuals and 100 generations has: 10,000, 10,000, 20,000, 40,000, 80,000, etc. For the mentioned algorithms, the numbers of steps divided by nominal numbers of steps gives equal numbers in all segments, for example for a single path 1, 1, 1, 1, 1, 1,… . Exemplary segments from 2-vagabonds verification of 3 philosophers in server deadlock verification contain 3, 5, 7, 19, 41, 23, 9 steps in individual segments. These values divided by nominal numbers of steps (1, 1, 2, 4, 8, 16, 32) give 3, 5, 4, 5, 5, 1, 0 (rounded to integers). Taking the number of steps in the first segment as 1, we get numbers of steps in consecutive as segments shown in Fig. 11.2, separately for 1st vagabond, absolute paths of 2nd vagabond and relative paths for 2nd vagabond. The absolute path consists of a subpath from the initial configuration T_0 to starting configuration T_{start} of 2nd vagabond plus the own path of 2nd vagabond. The relative path is the own 2nd vagabond path. The plots show that in some cases the numbers of steps decrease: this results from many short loops. In other cases, the assumption of avoiding frequent investigation of leader subpaths is fulfilled (the plots grow, than fall, or grow to their end). Plots for absolute path in 2nd vagabond grow rapidly, sometimes up to thousands.

Verifications of server deadlock (Servers dd), server-deadlock free (Servers not dd) and agent deadlock (Agent dd) are presented for 3, 10 and 15 philosophers in Fig. 11.2.

Table 11.1 Deadlock detection

Model	Type	Exhaustive search			2-vagabonds			
		Configuration space (thousands)	Space creation time (h:m:s)	Verif. Time for 1 server/ agent (h:m:s)	$D * N$ (thousands)	Avg steps in verification 1st vagabond (thousands)	Avg steps in verification 2nd vagabond (thousands)	Verif. Time for 1 server/ agent (h:m:s)
1	Servers	0.0	1	0	1.7	1.3	0.4	0
1	Agents	0.0	1	0	1.2	0.1	1.2	1
1	Agents t	0.0	1	0	1.2	1.2	0.2	1
2sem-cross	Servers	1.2	1	0	6.3	0.1	6.3	1
2sem-cross	Servers t	1.2	1	0	6.3	0.1	6.3	1
2sem-cross	Agents	1.2	1	0	4.6	0.1	4.6	1
2sem-cross	Agents t	1.2	1	0	4.6	0.1	4.6	2
Ph3	Servers	1.2	1	0	27.4	0.1	0.3	1
Ph3	Servers t	1.2	1	0	27.4	0.4	0.2	0
Ph3	Agents	1.2	1	0	17.8	0.0	0.0	0
Ph3	Agents t	1.2	1	0	17.8	0.2	0.2	1
Ph4	Servers	12.7	17	0	47.4	0.5	3.4	6
Ph4	Servers t	12.7	17	0	47.4	0.9	0.6	1
Ph4	Agents	12.7	17	0	31.0	0.2	1.5	3
Ph4	Agents t	12.7	17	0	31.0	0.1	0.0	0
Ph5	Servers	135.2	1:06:07	2	150.5	0.2	0.8	1
Ph5	Servers t	135.2	1:06:07	0	150.5	0.3	0.2	1
Ph5	Agents	135.2	1:06:07	4	100.3	0.2	2.3	3
Ph5	Agents t	135.2	1:06:07	0	100.3	0.2	0.5	0
Ph10	Servers				988.0	5.7	241.7	1:05

(continued)

Table 11.1 (continued)

Model	Type	Exhaustive search			2-vagabonds			
		Configuration space (thousands)	Space creation time (h:m:s)	Verif. Time for 1 server/ agent (h:m:s)	$D * N$ (thousands)	Avg steps in verification 1st vagabond (thousands)	Avg steps in verification 2nd vagabond (thousands)	Verif. Time for 1 server/ agent (h:m:s)
Ph10	Servers t				988.0	26.9	12.2	12
Ph10	Agents				653.6	6.7	660.2	3:30
Ph10	Agents t				653.6	60.5	29.4	23
Ph15	Servers				2211.6	14.8	2258.8	2:34:44
Ph15	Agents				1459.2	32.0	6091.4	31:50
Ph18	Servers				4801.7	36.3	9245.4	12:56:28
Ph18	Agents				3160.1	15.2	1532.0	3:22:08
PC5open	Servers				576.0	0.2	0.3	1
PC5open	Agents				489.6	0.2	0.3	1
PC10open	Servers				1364.2	0.4	0.5	1
PC10open	Agents				1698.2	0.4	0.5	3
PC15open	Servers				2786.4	0.7	0.6	1
PC15open	Agents				4540.8	0.7	0.7	7
Cross-straight	Servers	47.4	4:42	0	794.1	0.1	0.1	0
Cross-straight	Agents	47.4	4:42	0	506.9	0.2	0.2	1
Cross-all	Servers				865.5	0.3	0.1	1
Cross-all	Agents				547.6	0.8	1.7	6
AVGS	Servers	0.0	3	0	5.8	0.0	0.0	0
AVGS	Agents	0.0	3	0	4.5	0.0	0.0	1
Buf-3 × 3	Servers	0.5	0	0	5.3	0.0	0.0	0

(continued)

Table 11.1 (continued)

Model	Type	Exhaustive search			2-vagabonds			
		Configuration space (thousands)	Space creation time (h:m:s)	Verif. Time for 1 server/ agent (h:m:s)	$D * N$ (thousands)	Avg steps in verification 1st vagabond (thousands)	Avg steps in verification 2nd vagabond (thousands)	Verif. Time for 1 server/ agent (h:m:s)
Buf-3 × 3	Agents	0.5	0	0	4.8	0.0	0.0	0
Buf-6 × 6	Servers	109.4	1:08:48	0	65.7	0.3	0.1	0
Buf-6 × 6	Agents	109.4	1:08:48	0	82.9	0.2	0.0	0
Buf-10 × 10	Servers				430.1	0.2	0.0	1
Buf-10 × 10	Agents				706.6	0.6	0.2	2

Table 11.2 Safety from deadlock

Model	Type	Exhaustive search			2-vagabonds		
		Configuration space (thousands)	Space creation time (h:m:s)	Verification time for 1 server/agent (h:m:s)	$D * N$ = steps in verification 1st vagabond (thousands)	Avg steps in verification 2nd vagabond (thousands)	Verification time for 1 server/agent (h:m:s)
1	Servers	0.0	1	0	1.7	0.0	0
2sem-agr	Servers	0.1	1	0	6.3	0.0	0
Ph3	Servers	1.2	1	0	27.4	0.0	1
Ph4	Servers	12.7	17	0	47.4	0.0	2
Ph5	Servers	135.2	1:06:07	0	150.5	0.0	11
Ph10	Servers				988.0	0.0	4:15
Ph15	Servers				2211.6	0.0	14:49
Ph3-2orNo	Servers	3.2	4	0	81.4	64.8	5:34
Ph3-2orNo	Agents	3.2	4	0	44.9	145.0	11:21
Ph3-asym	Servers	0.6	0	0	22.8	24.3	1:05
Ph3-asym	Agents	0.6	0	0	15.6	39.6	2:05
Ph3-butlers	Servers	10.3	10	0	66.8	13.2	27
Ph3-butlers	Agents	10.3	10	0	34.6	28.8	1:04
Ph4-2orNo	Servers	31.5	2:46	0	184.7	123.8	9:40
Ph4-2orNo	Agents	31.5	2:46	0	109.8	455.6	32:25
Ph4-asym	Servers	6.1	3	0	40.8	80.8	2:48
Ph4-asym	Agents	6.1	3	0	27.7	159.4	2:46
Ph4-butlers	Servers	250.3	2:58:50	0	175.1	44.0	1:24
Ph4-butlers	Agents	250.3	2:58:50	0	96.8	102.7	3:41
Ph5-2orNo	Servers	295.0	4:48:03	31	430.6	243.2	18:44

(continued)

Table 11.2 (continued)

Model	Type	Exhaustive search			2-vagabonds		
		Configuration space (thousands)	Space creation time (h:m:s)	Verification time for 1 server/agent (h:m:s)	$D * N$ = steps in verification 1st vagabond (thousands)	Avg steps in verification 2nd vagabond (thousands)	Verification time for 1 server/agent (h:m:s)
Ph4-2orNo	Agents	295.0	4:48:03	19	252.7	1183.1	1:21:49
Ph5-asym	Servers	63.6	11:14	0	86.7	738.0	13:02
Ph5-asym	Agents	63.6	11:14	1	86.7	722.5	12:43
Ph5-butlers	Servers				441.6	159.8	5:32
Ph5-butlers	Agents				259.2	362.9	10:10
PC3	Servers	117.5	1:03:04	1	282.6	1108.2	29:39
PC3	Agents	117.5	1:03:04	44	273.8	3899.9	1:59:29
PC4	Servers				431.1	23.4	47
PC4	Agents				260.3	2036.5	33:16
PC5	Servers				588.8	1285.3	25:14
PC5	Agents				515.2	4736.5	1:12:09
Cross-right	Servers	74.0	18:21	0	794.1	73.2	10:0
Cross-right	Agents	74.0	18:21	0	506.9	405.8	18:11
AVGS-goBack	Servers	0.4	3	0	112.5	1.1	23
AVGS-goBack	Agents	0.4	3	0	51.1	15.6	1:19
AVGS-turnout	Servers	0.1	0	0	12.7	7.7	4
AVGS-turnout	Agents	0.1	0	0	7.7	9.2	29
AVGS-twoWays	Servers	0.2	0	0	17.6	0.2	3
AVGS-twoWays	Agents	0.2	0	0	11.1	3.4	17

Table 11.3 Termination

Model	Yes/not	Exhaustive search			2-vagabonds	
		Configuration space (thousands)	Space creation time (h:m:s)	Verification time for 1 server/agent (h:m:s)	$D * N$ = steps in verification 1st vagabond (thousands)	Verification time for 1 server/agent (h:m:s)
1	Yes/not	0.0	1	0	1.2	0
2sem-agr	Yes	0.1	1	0	4.6	28
Ph3	Not	1.2	1	0	17.8	1
Ph4	Not	12.7	17	1	31.0	3
Ph5	Not	135.2	1:06:07	2:29	100.3	16
Ph10	Not				653.6	1:44
Ph15	Not				1459.2	47:12
Ph3-2orNo	Not				44.9	11:22
Ph3-asym	Not				15.6	1:54
Ph3-butlers	Not				34.6	1:07
Ph4-2orNo	Not				109.8	32:25
Ph4-asym	Not				27.7	4:11
Ph4-butlers	Not				96.8	3:42
Ph5-2orNo	Not				86.7	1:21:58
Ph5-asym	Not				109.8	12:53
Ph5-butlers	Not				259.2	10:16
AVGS	Not	0.0	3	0	11.1	0
AVGS-goBack	Yes	0.4	3	0	51.1	20
AVGS-turnout	Yes	0.1	0	0	7.7	3
AVGS-twoWays	Yes	0.2	0	0	11.1	16
Buf-3 × 3	Not	0.5	0	0	4.8	0
Buf-6 × 6	Not	109.4	1:08:48	60	82.9	0
Buf-i0 × 10	Not				706.6	2

11.4 Parallel Verification

In sequential implementation, the 1st vagabond starts the 2nd one in certain situations, and waits for the result of 2nd vagabond run, then continues or cuts 1st vagabond progress depending on the results of 1st and 2nd vagabond search. However, various types of concurrent computations can be applied.

- First and second vagabond can run in parallel. When the second vagabond search is to be launched, and the previous one is not finished yet, the first vagabond is suspended.
- Alternatively for suspending first vagabond, a buffer between first and second vagabond can be established. This prevents the second vagabond from starting

Table 11.4 Student exercises

Student	Type	Result	$D * N$	Steps in verification 1st vagabond (thousands)	Steps in verification 2nd vagabond (thousands)	Verification time for 1 server/agent (h:m:s)
Stud 1	Servers dd	Partial dd	787.3	393.8	1721.0	16:47
Stud 1	Agents dd	Total dd	387.0	0.2	1.8	2
Stud 2	Servers dd	Partial dd	5888.0	3728.0	0.9	18:10
Stud 2	Agents dd	Total dd	4416.0	0.2	1.8	3
Stud 3	Servers dd	Safe	988.0	988.0	3728.0	46:41
Stud 3	Agents dd	Safe	4557.8	557.8	6810.1	1:29:57
Stud 4	Servers dd	Safe	930.8	930.8	33 0306.3	23:46:39
Stud 4	Agents dd	Safe	160.0	160.0	732.9	15:13

frequently, which is expensive (starting a new process or thread). The solution with a buffer between one 1st type vagabond and one 2nd type vagabond is used in our implementation of the algorithm in the Dedan program.

- An important feature of the 2-vagabonds algorithm is that multiple vagabonds of 1st and 2nd type can be launched. They run independently, except that finding a deadlock (or lack of termination) by a single vagabond should stop the run of all vagabonds. Of course, a number of vagabonds highly exceeding the number of available processors is useless.
- The configurations in the search path are elaborated on the run, therefore the input of the algorithm consists only of the definition of a set of actions. The input can be easily spread over a set of distributed computers to run several vagabonds in parallel. Both 1st and 2nd type vagabonds can be started in distributed way. Again, the whole set of vagabonds must be stopped after identifying the deadlock (or lack of termination). This solution is planned as the next step of the development of Dedan.

11.5 Limitations

The exhaustive search is proof of the verified feature. A non-exhaustive search may, in its nature, leave given features unverified. The methods mentioned in related work (Sect. 11.1) leave some uncertainty. If we treat finding a deadlock or a distributed termination as a positive result, false positive cannot happen. If a deadlock or termination is found, it is proved. However, a false negative may happen, i.e., existing deadlock or termination may not be found. We never have

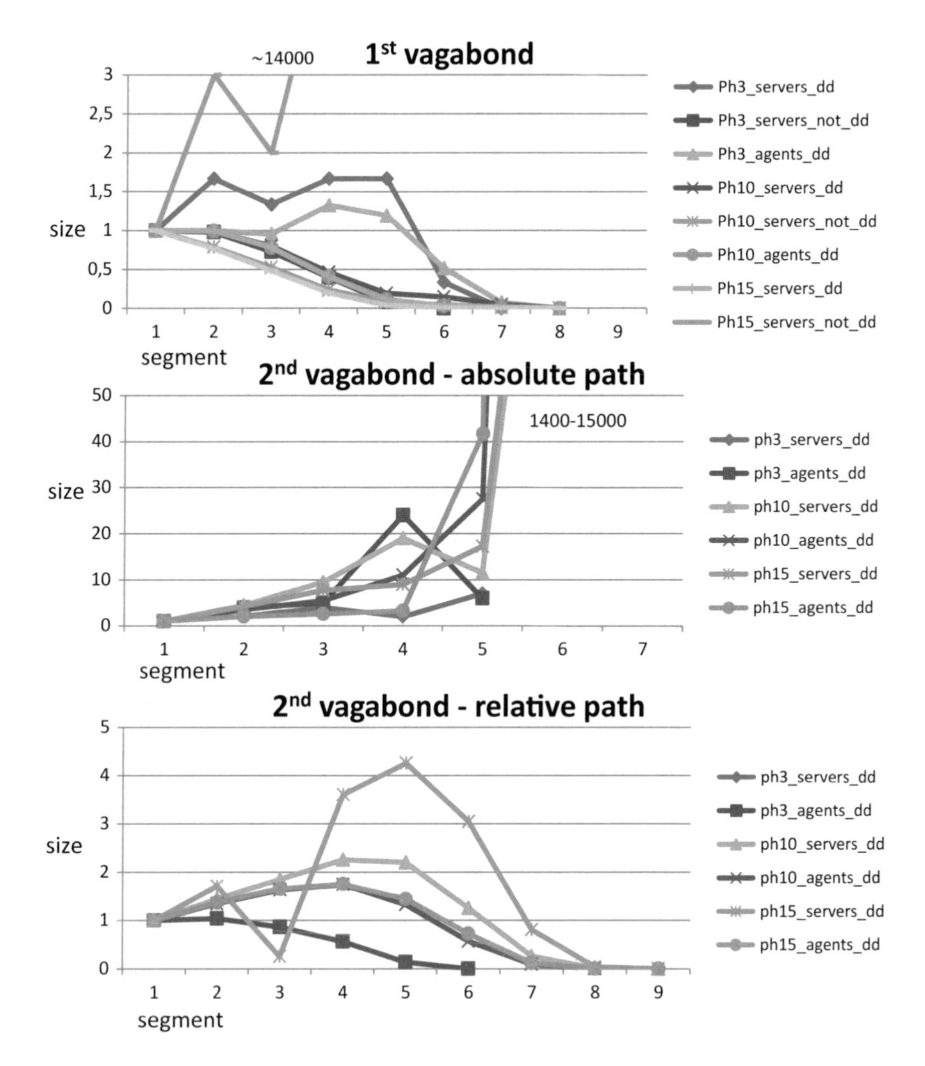

Fig. 11.2 Number of steps in individual segments in various verifications

proof that the entire reachability space is searched, even if we tested more states than fully connected graph of possible configurations:

$$N_{\max} = \prod_{k=1}^{n} card(\Lambda s_k) * \prod_{k=1}^{m} card(\Lambda a_k) \tag{11.2}$$

If we want to be sure that all reachable configurations are searched, we must build a full reachability space and mark every visited configuration. In the case of a partial deadlock and partial termination, we can also be sure that every reachable transition (an action between a given pair of configurations) is visited. This makes heuristic search senseless.

The search for partial features (deadlock and termination) gives another dimension of uncertainty: false positive is also possible. Recall that in the deadlock detection, after finding a configuration if which a process is not enabled, we search for a configuration in which it is enabled in the future. If we fail to find such a configuration, the decision is made on deadlock freeness of this fragment of reachability space (and the search is continued if $D * N$ is not reached).

For an exhaustive search, the memory size is the main limitation. We can use reduction techniques, for example symbolic model checking, compositional model checking, abstraction techniques, bounded model checking, etc., We can also use partial order reduction. All those treatments, as well as increasing the memory size, shift the size limit of the verified models, but it always exists. The time of building the reachability space is also important, but one can say that we only need to wait enough time.

In the non-exhaustive model checking, we switch memory limits to time limits. The question is: how long can we wait for the result? The longer we wait, the greater is our sureness that the result is right. To get more confidence in the result, we should extend the verification time. Look at deadlock detection in the case of philosophers problem: we tested a partial deadlock for every process (server or agent) in the sequence. We can ask for searching a more general deadlock, in which all the forks participate (chairs are not prone to deadlock). Or, alternatively, we can test the total deadlock of all agents (philosophers). In our formalism, the total deadlock (or deadlock concerning all processes of a given type, such as for the forks) is simply a partial deadlock in which a set of processes take part (possibly all).

Surprisingly, the verification of common deadlock of all forks or all philosophers gives a false negative result (deadlock freeness is reported) for 15 philosophers. This is due to the fact that finding a configuration in which all processes are disabled is easier when the set is smaller. We can comfort ourselves that finding a partial deadlock is an important result, because each type of deadlock should be analyzed and repaired (both partial and total). We can also trust that by eliminating the partial deadlock we also remove the possible deadlock comprising more processes, maybe a total deadlock. Otherwise, we can increase the factor D. However, hasty changing of D should be avoided because it affects the verification time, mainly in deadlock-free examples.

We must remember that if multiple processes fall in partial single-process deadlocks, they not necessarily fall into a common deadlock. This concerns especially deadlocks owing from lack of resources, as in example "1". This example shows that every agent is prone to deadlock, but never together.

In our algorithm we use the function for N (Eq. 11.1) which grows much slower than N_{max} (Eq. 11.2), due to the use of logarithms. The use of logarithms results

Table 11.5 Dijsktra's philosophers in comparison of non-exhaustive search methods

Paper	Modeling language	Technique	No. of philosophers	Verification time (h:m:s)
Francesca et al. (2011)	CCS	ACO	15	?
Francesca et al. (2011)	CCS	A*	12	?
Godefroid and Khurshid (2002)	C	GA	(52% runs) 17	2:57
Godefroid and Khurshid (2002)	C	GA (mutation only)	(52% runs) 17	2:16
Gradara et al. (2006)	CCS	A*	12	?
Gradara et al. (2006)	CCS	Greedy	40	?
Alba and Chicano (2007)	Promela	ACO	8	1:12:14
Yousefian et al. (2014)	GTS[a]	GA	30	40:09
Yousefian et al. (2014)	GTS[a]	A*	8	24:37
Pira et al. (2017)	GTS[a]	GA + machine learning	100	56
Pira et al. (2017)	GTS[a]	GA	20	23
Pira et al. (2017)	GTS[a]	A*	10	3:57
This monograph	IMDS	2-vagabonds	18[b]	84:18:51
This monograph	IMDS	2-vagabonds	12[c]	6:38:53

[a]GTS—Graph Transformation System
[b]Partial deadlock of one philosopher, obviously participating in total deadlock
[c]Total deadlock = partial deadlock of all processes

from the assumption that in a set of many processes, the dependencies between them do not form a large network of relationships. It is evident, however, that in some cases this function would give to small N, because generally N_{max} grows exponentially with number of servers and agents, while N grows logarithmically. Therefore, we should increase the D factor to be more certain about verification results. We must also remember that finding partial deadlocks gives a more reliable result.

The time complexity of the proposed 2-vagabonds algorithm (for the sequential version) depends on N and the factor D. For 1 vagabond (total deadlock detection) it is $O(D * N)$. For partial deadlock detection it is $O((D * N)^2)$, because for every configuration of the 1st search, the 2nd vagabond can search $D * N$ configurations. For termination detection, the last search for a terminating action may exceed $D * N$ with one more $D * N$, and the second vagabond searches similarly to deadlock detection, so the complexity is again $O((D * N)^2)$ (factor 2 is rejected).

It is difficult to compare results with other algorithms. First of all, the IMDS formalism is addressed to distributed systems, which is why each system must be built from independent servers communicating via messages. The global state of the system is not available for individual servers making their own decisions. Such benchmarks as Solitaire Game (Gradara et al. 2006), mentioned in many papers, cannot be examined in IMDS formalism. Another commonly used example—philosophers (Dijkstra 1959)—must be rebuilt to the form of a distributed system: *forks* are distributed servers, and philosophers must start from some servers, which is why a set of new servers called *chairs* is added (Fig. 5.6). Such a model ceases to be directly comparable, because its structure is different.

Second: our algorithm is addressed to detect partial deadlock/termination. Total deadlock/ termination can be identified, but in the sense of a partial deadlock/ termination, which comprises a set of processes, giving the full set. In the case of other algorithms, only total deadlock is examined due to their nature.

Third: all benchmarks used in the mentioned papers (Francesca et al. 2011; Godefroid and Khurshid 2002; Gradara et al. 2006; Chicano and Alba 2008; Alba and Chicano 2007; Yousefian et al. 2014; Pira et al. 2017; and others) show how quickly a deadlock can be found in the system that is known for its falling into a deadlock. Systems known to be safe or systems unknown to be safe/unsafe are not tested. Such an approach is appropriate for evaluation of the algorithm and the verification software. However, in practice, we need deadlock-free systems, proven by a model checker, rather than confirming that a large number of philosophers (say, 1000) fall into a deadlock. None of the above papers describes how long it takes to judge that a verified system is safe from deadlock. Such a comparison would be very interesting when comparing various non-exhaustive algorithms.

Table 11.5 compares various heuristic-based methods of deadlock detection with their maximum number of surveyed philosophers.

References

Alba, E., & Chicano, F. (2007). Ant colony optimization for model checking. In R. Moreno Díaz, F. Pichler, & A. Quesada Arencibia (Eds.), *International Conference on Computer Aided Systems Theory—EUROCAST 2007, Las Palmas, Spain*, LNCS (Vol. 4739, pp. 523–530), 12–16 February, 2007. Berlin Heidelberg: Springer. https://doi.org/10.1007/978-3-540-75867-9_66.

Alba, E., & Troya, J. M. (1996). Genetic algorithms for protocol validation. In H.-M. Voigt, W. Ebeling, I. Rechenberg, & H.-P. Schwefel (Eds.), *International Conference on Parallel Problem Solving from Nature PPSN 1996—PPSN IV, Berlin, Germany* (pp. 869–879), 22–26 September, 1996. Berlin, Heidelberg: Springer. https://doi.org/10.1007/3-540-61723-x_1050.

Batista, G. E. A. P. A., Prati, R. C., & Monard, M. C. (2004). A study of the behavior of several methods for balancing machine learning training data. *ACM SIGKDD Explorations Newsletter, 6*(1), 20–29. https://doi.org/10.1145/1007730.1007735.

Chicano, F., & Alba, E. (2008). Ant colony optimization with partial order reduction for discovering safety property violations in concurrent models. *Information Processing Letters, 106*(6), 221–231. https://doi.org/10.1016/j.ipl.2007.11.015.

Clarke, E. M., Long, D. E., & McMillan, K. L. (1989). Compositional model checking. In *Fourth Annual Symposium on Logic in Computer Science, Pacific Grove, CA* (pp. 353–362), 5–8 June 1989. IEEE Computer Society Press. https://doi.org/10.1109/lics.1989.39190.

Clarke, E. M., Grumberg, O., & Long, D. E. (1994). Model checking and abstraction. *ACM Transactions on Programming Languages and Systems, 16*(5), 1512–1542. https://doi.org/10.1145/186025.186051.

Clarke, E., McMillan, K., Campos, S., & Hartonas-Garmhausen, V. (1996). Symbolic model checking. In R. Alur & T. A. Henzinger (Eds.), *8th International Conference on Computer Aided Verification—CAV'96, New Brunswick, NJ* (pp. 419–422), 31 July–3 August, 1996. Berlin, Heidelberg: Springer. https://doi.org/10.1007/3-540-61474-5_93.

Czejdo, B., Bhattacharya, S., Baszun, M., & Daszczuk, W. B. (2016). Improving resilience of autonomous moving platforms by real-time analysis of their cooperation. *Autobusy-TEST, 17* (6), 1294–1301. url: http://www.autobusy-test.com.pl/images/stories/Do_pobrania/2016/nr%206/logistyka/10_l_czejdo_bhattacharya_baszun_daszczuk.pdf.

Daszczuk, W. B. (2018a). Siphon-based deadlock detection in integrated model of distributed systems (IMDS). In *Federated Conference on Computer Science and Information Systems, 3rd Workshop on Constraint Programming and Operation Research Applications (CPORA'18), Poznań, Poland* (pp. 421–431), 9–12 September, 2018. IEEE. https://doi.org/10.15439/2018f114.

Daszczuk W B. (2019a). Asynchronous specification of production cell benchmark in integrated model of distributed systems. In R. Bembenik, L. Skonieczny, G. Protaziuk, M. Kryszkiewicz, & H. Rybinski (Eds.), *23rd International Symposium on Methodologies for Intelligent Systems, ISMIS 2017, Warsaw, Poland, Studies in Big Data* (Vol. 40, pp. 115–129), 26–29 June, 2017. Cham, Switzerland: Springer International Publishing. https://doi.org/10.1007/978-3-319-77604-0_9.

Daszczuk, W. B. (2019b). Fairness in temporal verification of distributed systems. In W. Zamojski, J. Mazurkiewicz, J. Sugier, T. Walkowiak, & J. Kacprzyk (Eds.), *13th International Conference on Dependability and Complex Systems DepCoS-RELCOMEX, Brunów, Poland, AISC* (Vol. 761, pp. 135–150), 2–6 July, 2018. Cham, Switzerland: Springer International Publishing. https://doi.org/10.1007/978-3-319-91446-6_14.

Dijkstra, E. W. (1959). A note on two problems in connexion with graphs. *Numerische Mathematik, 1*(1), 269–271. https://doi.org/10.1007/BF01386390.

Francesca, G., Santone, A., Vaglini G., & Villani, M. L. (2011). Ant colony optimization for deadlock detection in concurrent systems. In *2011 IEEE 35th Annual Computer Software and Applications Conference, Munich, Germany* (pp. 108–117), 18–22 July, 2011. IEEE. https://doi.org/10.1109/compsac.2011.22.

Godefroid, P., & Khurshid, S. (2002). Exploring Very Large State Spaces Using Genetic Algorithms. In J.-P. Katoen & P. Stevens (Eds.), *International Conference on Tools and Algorithms for the Construction and Analysis of Systems—TACAS 2002, Grenoble, France, LNCS* (Vol. 2280, pp. 266–280), 8–12 April, 2002. Berlin Heidelberg: Springer. https://doi.org/10.1007/3-540-46002-0_19.

Gradara, S., Santone, A., & Villani, M. L. (2006). DELFIN+: An efficient deadlock detection tool for CCS processes. *Journal of Computer and System Sciences, 72*(8), 1397–1412. https://doi.org/10.1016/j.jcss.2006.03.003.

Hatcliff, J., Dwyer, M. B., & Zheng, H. (2000). Slicing software for model construction. *Higher-Order and Symbolic Computation, 13*(4), 315–353. https://doi.org/10.1023/A:1026599015809.

Jard C, and Jéron T. (1992). Bounded-memory algorithms for verification on-the-fly. In *International Conference on Computer Aided Verification—CAV 1991, Aalborg, Denmark* (pp. 192–202), 1–4 July, 1991. Berlin Heidelberg: Springer. https://doi.org/10.1007/3-540-55179-4_19.

Milner, R. (1984). *A calculus of communicating systems* (Vol. 92, LNCS). Berlin, Heidelberg: Springer. https://doi.org/10.1007/3-540-10235-3

Penczek, W., Gerth, R., Kuiper, R., Szreter, M. (1999). Partial order reductions preserving simulations. In *Concurrency Specification and Programming (CS&P), Warsaw, Poland* (pp. 153–171), 28–30 September, 1999. url: http://content.iospress.com/articles/fundamenta-informaticae/fi43-1-4-13.

Pira, E., Rafe, V., & Nikanjam, A. (2017). Deadlock detection in complex software systems specified through graph transformation using Bayesian optimization algorithm. *Journal of Systems and Software*, *131*, 181–200. https://doi.org/10.1016/j.jss.2017.05.128.

Santone, A. (2002). Automatic verification of concurrent systems using a formula-based compositional approach. *Acta Informatica, 38*(8), 531–564. https://doi.org/10.1007/s00236-002-0084-5.

Stirling, C., & Walker, D. (1991). Local model checking in the modal mu-calculus. *Theoretical Computer Science, 89*(1), 161–177. https://doi.org/10.1016/0304-3975(90)90110-4.

Yousefian, R., Rafe, V., & Rahmani, M. (2014). A heuristic solution for model checking graph transformation systems. *Applied Soft Computing*, *24*, 169–180. https://doi.org/10.1016/j.asoc.2014.06.055.

Chapter 12
Conclusions and Further Work

Communication duality, expressed as the server view and the agent view of the verified system, is the sound effect of the author's work. In the specification, it is simply grouping of actions in servers or in agents. The two views are the system decompositions that show message passing and resource sharing aspects of a distributed system. Figure 12.1 presents the metaphor of the decompositions. In the server view, an observer sees flowers (servers) with a set of bees flying between them and transferring pollen portions (messages). In the agent view, the observer focuses on individual bees traveling between flowers. The application of IMDS formalism highlights locality, autonomy and asynchrony in distributed systems, and distinguishes between features relative to agents behavior and features relative to servers behavior. The author claims that asynchronous specification is better suited for distributed systems than many synchronous modeling formalisms. The IMDS model, introduced in Chap. 3, combines asynchrony with communication duality. The specification formalism that supports communication duality, locality, autonomy and asynchrony in distributed systems, is an important research contribution of the monograph.

The IMDS establishes a global system configuration for building the Labeled Transition System and to perform model checking. However, in the definition of actions, and in the IMDS specification language, only local state of the server and messages pending at this server are observed. The only way to influence the behavior of the server is to send a message to it, and there is no external control over the server. It autonomously decides which of the pending messages will be accepted and when. Therefore, the formalism is especially tailored to real distributed systems, like Local Area Networks (LANs), internet services, multiagent systems, etc. Modeling truly distributed systems is based on simple protocols between servers, rather than on common variables. Such modeling is appropriate for Internet of Things approach (IoT, (Lee et al. 2013)), where independent devices autonomously negotiate their coordinated behavior using simple protocols. A few examples of systems specified this way are contained in Chap. 5.

© Springer Nature Switzerland AG 2020

W. B. Daszczuk, *Integrated Model of Distributed Systems*, Studies in Computational Intelligence 817, https://doi.org/10.1007/978-3-030-12835-7_12

Fig. 12.1 The two views: server view (flowers) and agent view (bees)

Distributed systems specification and verification formalisms typically identify total deadlocks, and sometimes total distributed termination. Some of them allow detection of partial deadlocks and/or termination, but at a expense of limiting the possible system structure or specifying the features manually by the user. An overview of those formalisms is given in Chap. 2. IMDS integrated with model checking provides automatic detection of partial and total deadlocks and partial and total termination by means of general temporal formulas. In addition, communication deadlocks can be identified in the server view while resource deadlocks in the agent view (Sect. 3.4 and Chap. 4). This feature is very rare among the known formalisms. The automatic verification of total/partial properties and differentiation between their kinds is the second main research contribution of the author.

The presented methodology is the basis of the Dedan specification and verification tool. Conversion between the views in specification, where the server/agent types and variables are introduced, is performed algorithmically in Dedan. The two views allow the observation of the system from the point of view of cooperating servers or from the point of view of traveling agents. In Sects. 5.3, 5.4 and 5.5, this is illustrated by examples of transportation systems with automatic vehicle guidance: the server view shows the perspective of cooperating road segment controllers, while the agent view—from the perspective of vehicles. While the controller view is natural to the computer engineer, the vehicle view might be more natural to the transport specialist. Counterexamples generated by Dedan show the message exchange between the controllers (server view) or vehicle movement using the segments (agent view). A resource deadlock without a communication deadlock—an agent which is stuck while all servers work—is presented in the "roundabout" example.

The Karlsruhe Production Cell benchmark was successfully modeled and verified under Dedan (Sect. 5.6). The verification of its timed version (using Uppaal) failed due to modeling asynchronous communication channels as separate Timed Automata. This largely extends the reachability space of the specification. Such verification should be possible in a future symbolic version of the internal Dedan verifier (explained below), because asynchrony of channels is modeled directly in the LTS of the IMDS specification. In addition, the verification using the timed "2-vagabonds" algorithm seems promising.

To check the deadlocks and termination, a designer should express the system (or—better—its model; or best—the communication skeleton of the system) in terms of IMDS elements: servers' states, agents' messages, and actions. The proposed notation is presented informally in Sect. 3.3, and defined formally in Appendix C. The verification process is carried automatically by the Dedan program, which finds communication deadlocks, resource deadlocks and agents termination. The user does not have to know any temporal logic to verify the model. Counterexamples and witnesses are generated automatically in a readable form and stored in various file formats. In addition, it is possible to simulate the verified system, and to observe the shape of the LTS. Several examples of distributed systems and their verification are presented in Chap. 5. The structure of the Dedan program and the overview of its use are shown in Chap. 6. The specification of general formulas for deadlock and termination verification (including partial ones) independently of specific features of a system, is a sound practical contribution.

Two specification formalisms are proposed, equivalent to IMDS: distributed automata DA^3 (in two forms, Chap. 8) and Petri nets with a restricted structure (Chap. 7). The equivalence of the two formalisms with IMDS lays in mapping their basic structures onto structures of IMDS and analogous reachability graphs: LTS for IMDS, reachable markings graph for a Petri net and global node space for distributed automata of both forms. Elements and verification features, wider than deadlock and termination identification, are collected for all three formalisms in Table 12.1. This variety of specification modes greatly facilitates the analysis of distributed system features.

Automation of deadlock detection and termination checking allows the students of ICS, WUT to use the presented methodology. They prepare their solutions to synchronization problems and then verify these solutions using Dedan. This procedure allows the students to analyze the correctness of their ideas, to correct wrong solutions, and to get used to formal verification of software. Other facilities of the Dedan program support this process: analysis of structural properties in a Petri net equivalent to the IMDS specification, and several modes of simulation of the system under verification. As an additional convenience, students may use other forms of specification, equivalent to IMDS: distributed automata DA^3 (described in Chap. 8) and the imperative-style Rybu preprocessor, prepared by the students if ICS, WUT under supervision of the author (Daszczuk et al. 2017).

Because external verifiers are used to verify large systems, the complexity of internal TempoRG verifier is not very important. Explicit reachability space generates known problems with the size of the space and the time of its elaboration

Table 12.1 Verification facilities in the three equivalent formalisms

Formalism	IMDS	Petri net	DA^3
Main features	Specification, model checking, simulation	Structural properties	Graphical input, simulation
Notions	State	"Red" place	• Node (S-DA^3) • Element of global input vector, input/output symbol on transitions (A–DA^3)
	Message	"Green" place	• Element of input set (S–DA^3), input/output symbol on transitions • Node (A-DA^3)
	Configuration	Marking	Global node
	Action	Transition	Transition
	Initial state	Token in red place in initial marking	• Initial node (S–DA^3) • Initial element of global input vector (A–DA^3)
	Initial message	Token in green place in initial marking	• Initial element of input set (S–DA^3) • Initial node (A–DA^3)
	Initial configuration	Initial marking	• Initial nodes and initial input sets of all automata (S–DA^3) • Initial nodes and initial global input vector (A–DA^3)
	Labeled transition system	Marking reachability graph	Global node space: • All states and input sets in global nodes, input and output symbols on transitions (S–DA^3) • All messages and global input vector in global nodes, input and output symbols on transitions (A–DA^3)
Features	• Resource deadlock • Communication deadlock • Partial deadlock • Total deadlock • Partial distributed termination • Total distributed termination • Counterexamples/witnesses • Configuration space inspection • Simulation over configuration space	• Structural properties • Unreachable actions • Separated components • Existence of siphons • Many deadlocks • Invariants	• Graphical definition of a system • Simulation over individual automata • Counterexample projected onto individual automata • Counterexample–guided simulation

(McMillan 1993). They depend more on the parallelism degree of the modeled system and the way of cooperation within the system than on the size of the specification. The "intersection" example with 4 roads, having 73 lines of source code, has a reachability space that exceeds the possibilities of internal verification. Its size is over 2 million configurations, but it can be checked using an external verifier. On the other hand, one of the students' solutions to synchronization problems, consisting of almost 6 thousand lines of source code (generated by the Rybu preprocessor), is successfully verified in about half a minute (Sect. 5.8). The reachability space contains almost 15 thousand configurations. For many internally verified systems, reachability space is generated in a few minutes and verification time is several seconds.

The Dedan verification environment is under development. In addition to TempoRG verifier, other model checkers are used, including Spin (Holzmann 1995, 1997), symbolic NuSMV model checker (Cimatti et al. 2000) and Uppaal (Behrmann et al. 2006). Model checking with TempoRG on explicit reachability space will be performed with space reduction, which will allow to verify larger systems without the usage of external verifiers (which sometimes show problems with fairness, described in Chap. 9). Typical reachability space reduction techniques are based on an independence relation, which is supported externally (Penczek et al. 2000; Godefroid and Wolper 1992; Gerth et al. 1999). IMDS is attractive in this matter, because the independence relation is natural: actions on separate servers are always independent. In addition, a symbolic representation of a reachability space in Binary Decision Diagrams (BDD) is planned (Clarke et al. 1999). This will allow verification of large systems entirely inside the program, without external verifiers.

The behavior of a distributed system can strongly depend on real time dependencies. Time constraints imposed on action execution or communication delay can change the behavior: some deadlocks can possibly disappear or new deadlocks may possibly arise. Besides, improper timing may prevent a successful distributed termination. Verification with real time constraints is performed externally, using Uppaal verifier (Behrmann et al. 2006), after conversion of the verified system to Timed Automata (Alur and Dill 1994; Hui and Chikkagoudar 2012) (Chap. 10). Time constraints can be imposed on time delays of communication channels, duration of actions and time limits of staying in the states. For example, this allows to model and verify urban transportation problems. In the future, discrete time verification will be also applied, using the specification in EmLan language (Krystosik 2006; Krystosik and Turlej 2006). The timed version of 2-vagabonds non-exhaustive verification algorithm is under tests. The timed specification and verification preserving communication duality, locality, autonomy and asynchrony is an important research and practical contribution.

Chapter 11 presents the non-exhaustive 2-vagabonds algorithm for partial deadlock and partial termination verification. The algorithm is tailored especially for distributed IoT systems consisting of a number on servers running a set of cooperating agents. The algorithm was tested on several examples, both falling into a deadlock and safe from deadlock. Some student exercises with a result unknown in advance were also examined.

The algorithm is the only one known which can non-exhaustively search for partial deadlock and termination. Its construction confirms our assumption: it has only one parameter D controlling the number of steps performed during heuristic search. Also, in many cases it searches leader subpaths less frequently than the rest of reachability space.

Both algorithms, CBS and 2-vagabonds, are designed for the IMDS specification, but we believe that they can be applied to other formalisms. However, IMDS formalism is the only one which differentiates deadlock from termination independently of the shape of the verified system, and which can identify communication deadlock in the server view and resource deadlock in the agent view (Daszczuk 2017a).

The 2-vagabonds algorithm uses two processes (vagabonds): first puts the hypothesis and the second verifies the hypothesis. Because they work almost independently, various parallel and distributed implementations are possible. Elements of other approaches: ACO, GA, A* and machine learning can increase the performance of 2-vagabonds. We plan to apply these heuristics in our future research. Partial order reductions can also be applied, especially because independence relation is built in the formalism: actions executed in separate servers are always independent (while actions in the same server are always in conflict).

The future directions of IMDS and Dedan development are:

- Reachability space reduction and symbolic representation, mentioned above.
- Checking for assertions expressed in terms of states and messages.
- Invariant discovery—relations between states and messages held in loops of LTS.
- Higher level language-based input—elaboration of two languages of distributed systems specification: for the server view—exploiting locality in servers and message passing, and for the agent view—exploiting traveling of agents and resource sharing in the distributed environment.
- Code mobility—equipping the agents with their own sets of actions, carried in their "backpacks", parameterizing their behavior. This will allow modeling of mobile agents and let avoiding the specification of multiple slightly differing server types. An example of such a specification is "dining philosophers" with asymmetric solution—some philosophers take their left forks first and some of them take right forks first.
- Dynamic creation of processes—a state of new server and/or a message of new agent created in the action (Chrobot and Daszczuk 2006). Some problems with the elaboration of the reachability space of such system must be solved first. To be statically verified, the LTS of a Petri net corresponding to the IMDS system should be *bounded*, e.g. by limiting markings to some bound (van der Aalst 1998).

- Non-exhaustive search for partial deadlock detection and partial termination checking in timed systems (with real-time constraints).
- Probabilistic model checking of models with probabilities of choices between the actions in servers and in agents (Kwiatkowska et al. 2011; Alexiou et al. 2016).
- Minor improvements, for example indexing of vectors from 0 and command-line interface to run Dedan from batch.

References

Alexiou, N., Basagiannis, S., & Petridou, S. (2016). Formal security analysis of near field communication using model checking. *Computers & Security, 60,* 1–14. https://doi.org/10.1016/j.cose.2016.03.002.

Alur, R., & Dill, D. L. (1994). A theory of timed automata. *Theoretical Computer Science, 126*(2), 183–235. https://doi.org/10.1016/0304-3975(94)90010-8.

Behrmann, G., David, A., & Larsen, K. G. (2006). *A tutorial on Uppaal 4.0*. Aalborg, Denmark. URL: http://www.it.uu.se/research/group/darts/papers/texts/new-tutorial.pdf.

Chrobot, S., & Daszczuk, W. B. (2006). Communication dualism in distributed systems with Petri net interpretation. *Theoretical and Applied Informatics, 18*(4), 261–278. URL: https://taai.iitis.pl/taai/article/view/250/taai-vol.18-no.4-pp.261.

Cimatti, A., Clarke, E., Giunchiglia, F., & Roveri, M. (2000). NUSMV: A new symbolic model checker. *International Journal on Software Tools for Technology Transfer, 2*(4), 410–425. https://doi.org/10.1007/s100090050046.

Clarke, E. M., Grumberg, O., & Peled, D. (1999). *Model checking*. Cambridge, MA: MIT Press. ISBN: 0-262-03270-8.

Daszczuk, W. B. (2017). Communication and resource deadlock analysis using IMDS formalism and model checking. *The Computer Journal, 60*(5), 729–750. https://doi.org/10.1093/comjnl/bxw099.

Daszczuk, W. B., Bielecki, M., & Michalski, J. (2017). Rybu: Imperative-style preprocessor for verification of distributed systems in the Dedan environment. In *KKIO'17—Software Engineering Conference, Rzeszów, Poland*, 14–16 September, 2017. Polish Information Processing Society. arXiv:1710.02722.

Godefroid P, and Wolper P. (1992). Using partial orders for the efficient verification of deadlock freedom and safety properties. In *3rd International Workshop, CAV'91, Aalborg, Denmark*, LNCS 575 (pp. 332–342), 1–4 July, 1991. Berlin, Heidelberg: Springer. https://doi.org/10.1007/3-540-55179-4_32.

Holzmann, G. J. (1995). Tutorial: Proving properties of concurrent systems with SPIN. In *6th International Conference on Concurrency Theory, CONCUR'95, Philadelphia, PA* (pp. 453–455), 21–24 August, 1995. Berlin, Heidelberg: Springer. https://doi.org/10.1007/3-540-60218-6_34.

Holzmann, G. J. (1997). The model checker SPIN. *IEEE Transactions on Software Engineering, 23*(5), 279–295. https://doi.org/10.1109/32.588521.

Hui, P., & Chikkagoudar, S. (2012). A formal model for real-time parallel computation. *Electronic Proceedings in Theoretical Computer Science, 105,* 39–55. https://doi.org/10.4204/EPTCS.105.4.

Krystosik, A. (2006). Embedded systems modeling language. In *2006 International Conference on Dependability of Computer Systems,, DepCos-RELCOMEX'06, Szklarska Poręba, Poland* (pp. 27–34), 25–27 May, 2006 . Berlin Heidelberg: Springer. https://doi.org/10.1109/depcos-relcomex.2006.21.

Krystosik, A., & Turlej, D. (2006). EMLAN: A language for model checking of embedded systems software. In *IFAC Workshop on Programmable Devices and Embedded Systems, Brno, Czech Republic* (pp. 126–131), 14–16 February, 2006. Elsevier Science. https://doi.org/10.1016/S1474-6670(17)30171-4.

Kwiatkowska, M., Norman, G., & Parker, D. (2011). PRISM 4.0: Verification of probabilistic real-time systems. In *23rd International Conference, CAV 2011, Snowbird, UT* (pp. 585–591), 14–20 July, 2011. Berlin, Heidelberg: Springer. https://doi.org/10.1007/978-3-642-22110-1_47.

Lee, G. M., Crespi, N., Choi, J. K., Boussard, M. (2013). Internet of things. In *Evolution of telecommunication services*. LNCS 7768 (pp. 257–282). Berlin, Heidelberg: Springer. https://doi.org/10.1007/978-3-642-41569-2_13.

McMillan, K. L. (1993). *Symbolic model checking*. Norwell, MA: Kluwer Academic Publishers. ISBN: 0792393805.

Penczek, W., Gerth, R., Kuiper, R., & Szreter, M. (1999). Partial order reductions preserving simulations. In *Concurrency Specification and Programming (CS&P), Warsaw, Poland* (pp. 153–171), 28–30 September, 1999.

Penczek, W., Szreter, M., Rob, G., & Kuiper, R. (2000). Improving partial order reductions for universal branching time properties. *Fundamenta Informaticae, 43*(1–4), 245–267. url: http://www.ipipan.waw.pl/~penczek/WPenczek/papersPS/IPI843-97.ps.gz.

van der Aalst, W. M. P. (1998). The application of Petri nets to workflow management. *Journal of Circuits, Systems and Computers, 08*(01), 21–66. https://doi.org/10.1142/S0218126698000043.

Appendix A
Acronyms, Shortcuts and Symbols Used in the Text

^	Exponent
2^y	Powerset induced by a set y
$\Diamond$	LTL operator: eventually
$\Box$	LTL operator: always
$\lfloor x \rfloor$	Integer part of positive x
A	Set of agents
A-DA^3	Agent DA^3
a, a_1, a'	Agents belonging to *A*
AE, AN, AS, AW	Approaching roads in "road intersection" example—cardinal directions used
AF	CTL operator—always on a path
AG	CTL operator—always on all paths
ANDL	Abstract Net Description Language—file format for Petri nets
b	Set of vectors of variable values in TA, enabling a transition
B(*s*)	Server process of server *s*
BDD	Binary Decision Diagrams
BMP	Graphical file format
BT-IMDS	Timed IMDS with state time bounds and action firing time restrictions
C(*a*)	Agent process of agent *a*
c	Hidden clock used for counting action durations and channel delays in T-IMDS
c_0, c_1, …	Clocks of TA
card(*X*)	Cardinality of set *X*
CBS	Checking By Spheres—the algorithm of temporal formula evaluation, invented by the author and described in his Ph.D. Thesis

© Springer Nature Switzerland AG 2020

W. B. Daszczuk, *Integrated Model of Distributed Systems*, Studies in Computational Intelligence 817, https://doi.org/10.1007/978-3-030-12835-7

CCS	Calculus of Communicating Systems (formalism invented by Milner)
ch!, *ch?*	Sending and receiving over TA channel *ch*
Charlie	Petri net analyzer
clk	Clock used for counting state time bounds and action firing time restrictions in T-IMDS
constr	Set of clock values subranges
CSM	Concurrent State Machines (formalism invented by Jerzy Mieścicki)
CSP	Communicating Sequential Processes (formalism invented by C.A.R. Hoare)
CT-IMDS	Timed IMDS with channel time delays
CTL	Computation Tree Logic
D_a	*True* in all configurations where a message of the agent *a* is pending
DA^3	(D-triple A or DA-cubed) Distributed Asynchronous and Autonomous Automata—a formalism equivalent to IMDS
dda(*a*)	Agent *a* is in deadlock
dds(*s*)	Server *s* is in deadlock
Dedan	The program for verification of distributed systems, developed by the author
Delfin+	Deadlock finder based on CCS, invented by Gradara, Santone and Villani
D	Multiplier of number of steps in 2-vagabonds algorithm
D_s	Atomic Boolean formula *true* in all configurations where at least one message is pending at the server *s*
E	Set of transitions of TA
E1, E2	Markers placed on upper edge of track in "automatic vehicle guidance system" example
E_a	Atomic Boolean formula *true* in all configurations where an action is prepared with a message of the agent *a*
EF	CTL operator—eventually on a path
EmLan	The language for specification of concurrent systems, using discrete time constraints, invented by Artur Krystosik
E_s	Atomic Boolean formula *true* in all configurations where at least one action is prepared at the server *s*
EX	CTL operator—next on a path
F	Function assigning new values to the variables in TA
F_a	Atomic Boolean formula *true* in all configurations where a terminating action is prepared with a message of the agent *a*
F_i	Set of transitions of ith server DA^3 automaton $\{(p_1, m/m', p_2)$ *or* $(p_1, m/, p_2)\}$

G_i	Set of transitions of ith agent DA[3] automaton $\{(m_1, p/p', m_2)$ *or* $(m_1, p/p', t_{\varepsilon i})\}$
$g_{\Lambda s}$	Function mapping actions of server s to time durations (subranges)
g_{chj}	Function mapping asynchronous subchannels to time delays (subranges)
H	Set of all items (states and messages) $P \cup M$
ICS	Institute of Computer Science
$idle(s)$	Server s is idle
IMDS	Integrated Model of Distributed Systems—the formalism described in the monograph
$J(l)$	Function mapping TA locations to server clock values subranges
$J_{chj}(s_k)$	Subchannel time invariant
$J_l(l)$	Set of clock valuations of location l in TA
$J_{ll}(l, l')$	Set of clock valuations of transition l, l' in TA
$J_{\Lambda s}(\lambda)$	Action time invariant in server s
JPG	Graphical file format
$J_{\Lambda s}$	Action invariant mapping server's s clock to a subrange
$l_0, l_1, \ldots$	Locations of TA (l_0—initial location)
L	Set of locations of TA
Lab	Set of labels of TA transitions
LANs	Local Area Networks
L_s	Set of locations of TA corresponding to server s
L_{ch}	Set of locations of asynchronous channel
LTL	Linear Temporal Logic
LTS	Labeled Transition System
M	Markers placed in the middle of track in "automatic vehicle guidance system" example
M	Set of messages
m, m_1, m'	Messages of the form (*agent*, *server*, *service*) = (a, s, r)
$M(a)$	Set of messages of agent a
$M(s)$	Set of messages directed to server s
m_{0i}	Initial node of *i*th agent DA[3] automaton (a_i, s_j, r_{0j})
M_i	Set of nodes of *i*th agent DA[3] automaton $\{(a_i, s_j, r_j)$ *or* $t_{\varepsilon i}\}$
$M_{ini} \subset M$	Set of initial messages
MPA	Message Passing Automata
N	Set of nodes of LTS
N	*Number of steps in 2-vagabonds algorithm*
n_S	Number of servers
n_A	Number of agents
O	Set of variables of TA
$\bar{O}$	Vector of values of variables of TA
$\bar{O}_0$	Initial vector of values of variables in TA

$\{\bar{O}\}$	Set of vectors of variable values in TA
$\{\bar{O}[x] = r\}$	Vectors in $\{\bar{O}\}$ containing a value r as xth element
p_{0i}	Initial node of ith server DA^3 automaton (s_i, v_{0si})
P_i	Set of nodes of ith server DA^3 automaton $\{(s_i, v)\}$
p, p_1, p'	Server states of the form (*server*, *value*) = (s, v)
p_i	Node of ith server DA^3 automaton (s_i, v)
$p_{s\lambda d}(p_s, \lambda, g_\Lambda(\lambda))$	Location of TA connected with the state p_s
P	Set of servers' states
$P(s)$	Set of states of serves s
$P_{ini} \subset P$	Set of initial states
PDA	Pushdown distributed automata
PDF	Graphical file format
P_i	Set of nodes of ith server DA^3 automaton $\{(s_i, v)\}$
p_i	Node of ith server DA^3 automaton (s_i, v)
PNG	Graphical file format
Q	Set of synchronous UTA channels implementing T-IMDS asynchronous channels
QNE,QNW,QSE,QSW	Quarters or crossing in "road intersection" example—cardinal directions used
QSCTL	Temporal logic: Computation Tree Logic (CTL) with state quantification and operators relative to component state machines, invented by the author and described in his Ph.D. Thesis
R	Set of services
r, r_1, r'	Services belonging to R
R1…R6	Resources used in "automatic vehicle guidance system" example
RAG	Resource Allocation Graph
RPC	Remote Procedure Call
Rybu	Imperative language-based preprocessor of the Dedan program
S	Set of servers
S-DA^3	Server DA^3
s, s_1, s'	Servers belonging to S
SMV	Symbolic Model Verifier—model checker
Spin	Model checker
TA	Timed automata
T-IMDS	Timed version of IMDS
TCTL	Timed Computation Tree Logic
TempoRG	The verifier developed by the author of this monograph, described in his Ph.D. Thesis
term(a)	Agent a terminates
T_{ini}	Initial configuration

$T_{inp}(\lambda)$	Input configuration of action λ
$T_{out}(\lambda)$	Output configuration of action λ
$T_{Ʉ}$, $T_{Ʉ}'$	Global node of agent DA[3] automata (m_i, Y; i = 1..n)
$T_{Ʉ0}$	Initial global node of agent DA[3] automata (m_{0i}, Y_0; i = 1..n)
$T_{Ƨ}$, $T_{Ƨ}'$	Global node of server DA[3] automata ((p_i, X_i))
$T_{Ƨ0}$	Initial global node of server DA[3] automata ((p_{0i}, X_{0i}))
u	Valuation of set of clock variables in TA
u_0	All clocks equal 0
UML	Unified Modeling Language
Uppaal	Model checker
UTA	Uppaal timed automata
V	Set of values
v, v_1, v'	Values belonging to V
W	Set of transitions of LTS
W	Set of values of TA variable O
WFG	Wait-For Graph
WUT	Warsaw University of Technology
X_{0i}	Initial input set of server DA[3] automaton, $X_{0i} \in 2$^$\{m = (a, s_i, r_i)\}$
X_i	Input set of server DA[3] automaton, $X_i \in 2$^$\{m = (a, s_i, r_i)\}$
Y_0	Initial global input vector of the set of agent DA[3] automata $Y_0[i] = [(s_i, v_{0i})]$
Y	Global input vector of the set of agent DA[3] automata (the same for all ʉ in the system) $Y[i] = (s_i, v_i)$
$Y[i]$	ith element of global input vector Y
$\bar{Y}_p$	Set of current locations of all UTA server automata
$\bar{Y}_{ch}$	Set of current locations of all UTA channel automata
Z	Set of clocks of TA
Δ	Succession relation
Λ	Set of actions $\{(p, m)\ \Lambda\ (p', m')\} \cup \{(p, m)\ \Lambda\ (p')\}$
Λa	Set of actions of agent a
Λs	Set of actions of server s
λ	Action belonging to Λ
τ	Action of TA, other than sending or receiving over a channel
Ʉ	Set of agent DA[3] automata
ʉ, $ʉ_i$	Agent DA[3] automaton
Ƨ	Set of server DA[3] automata
ƨ, $ƨ_i$	Server DA[3] automaton
➤	Transition relation in TA

$\succ_d$	Timed transition relation in TA
$\succ_?$	Channel input transition relation in TA
$\succ_!$	Channel output transition relation in TA
$\succ_\lambda$	Transition relation in TA corresponding to actions λ
$\succ_{ch}$	Transition relation in TA corresponding message transfer in channel *ch*

Appendix B
IMDS Formal Definition

$P = \{p_1, p_2, \ldots\}$	The finite set of *states* (of *servers*)
$M = \{m_1, m_2, \ldots\}$	The finite set of *messages* (of *agents*)
$\Lambda \subset (M \times P) \text{ x } (M \times P) \cup (M \times P) \text{ x } (P)$	The set of *actions*
$P_{ini} \subset P$	The set of *initial states*
$M_{ini} \subset M$	The set of *initial messages*
$H = P \cup M$	The set of *items*
$T_{ini} = P_{ini} \cup M_{ini}$	The *initial configuration*
$T \subset H$	The *configuration*
$\forall_{\lambda \in \Lambda}\ \lambda = ((m, p), (m', p'))\ T_{inp}(\lambda) \supset \{m, p\},\ T_{out}(\lambda) = T_{inp}(\lambda) \backslash \{m, p\} \cup \{m', p'\}$	The rule of obtaining $T_{out}(\lambda)$ from $T_{inp}(\lambda)$ for the action λ
$\forall_{\lambda \in \Lambda}\ \lambda = ((m, p), (p')),\ T_{inp}(\lambda) \supset \{m, p\},\ T_{out}(\lambda) = T_{inp}(\lambda) \backslash \{m, p\} \cup \{p'\}$	The rule for the agent-terminating action

$$\text{LTS} = \langle N, n_0, W \rangle,$$

where:

- N is the set nodes (configurations $\{T_0, T_1, \ldots\}$, $T_{ini} = T_0$);
- n_0 is the root, $n_0 \in N$ (configuration T_{ini});
- W is the set of directed labeled transitions, $W \subset N \times \Lambda \times N$, $W = \{(T_{inp}(\lambda), \lambda, T_{out}(\lambda)) | \lambda \in \Lambda\}$.

$S = \{s_1, s_2, \ldots\}$	Finite set of servers
$A = \{a_1, a_2, \ldots\}$	Finite set of agents
$P(s) = \{p_{1s}, p_{2s}, \ldots\},\ s \in S$	Finite set of states of server s
$M(a) = \{m_{1a}, m_{2a}, \ldots\},\ a \in A$	Finite set of messages of agent a
$M(s) = \{m_{1s}, m_{2s}, \ldots\},\ s \in S$	Finite set of messages directed to server s

© Springer Nature Switzerland AG 2020

W. B. Daszczuk, *Integrated Model of Distributed Systems*, Studies in Computational Intelligence 817, https://doi.org/10.1007/978-3-030-12835-7

$P = \cup_{s \in S} P(s), \forall_{s1,s2 \in S}\ s_1 \neq s_2 \Rightarrow P(s_1) \cap P(s_2) = \varnothing$ — Every state is attributed to some server

$M = \cup_{a \in A} M(a) = \cup_{s \in S} M(s), \forall_{a1,a2 \in A}\ a_1 \neq a_2 \Rightarrow M(a_1) \cap M(a_2) = \varnothing, \forall_{s1,s2 \in S}\ s_1 \neq s_2 \Rightarrow M(s_1) \cap M(s_2) = \varnothing$ — Every message is attributed to some agent and is directed to some server

$\forall_{\lambda \in \Lambda}\ \lambda = ((m, p), (m', p')), m \in M(a) \Rightarrow m' \in M(a), p \in P(s) \Rightarrow p' \in P(s) \wedge m \in M(s)$ — Constraints on input and output pairs of the action

$B(s) = \{\lambda \in \Lambda | \lambda = ((p, m), (p', m')) \vee \lambda = ((p, m), (p')), p \in P(s)\}$ — The server process of the server *s*

$C(a) = \{\lambda \in \Lambda | \lambda = ((p, m), (p', m')) \vee \lambda = ((p, m), (p')), m \in M(a)\}$ — The agent process of the agent *a*

$\mathbf{B} = \{B(s) | s \in S\}$ — The server view

$\mathbf{C} = \{C(a) | a \in A\}$ — The agent view

IMDS as the programming language:

$S = \{s_1, s_2, \ldots\}$ — The finite set of servers

$A = \{a_1, a_2, \ldots\}$ — The finite set of agents

$V = \{v_1, v_2, \ldots\}$ — The finite set of values

$R = \{r_1, r_2, \ldots\}$ — The finite set of services

$P \subset S \times V$ — The set of states

$M \subset A \times S \times R$ — The set of messages

$$\Lambda \subset (M \times P) \times (M \times P) \cup (M \times P) \times (P) | (m,p)\Lambda(m',p') \vee (m,p)\Lambda(p'),$$
$$m = (a,s,r) \in M, p = (s_1,v_1) \in P, m' = (a_2,s_2,r_2) \in M,$$
$$p' = (s_3,v_3) \in P, s_1 = s, s_3 = s, a_2 = a$$

$B(s) = \{\lambda \in \Lambda | \lambda = (((a, s, r), (s, v)), ((a, s', r'), (s, v'))) \vee \lambda = (((a, s, r), (s, v)), ((s, v'))), s' \in S, a \in A, v, v' \in V, r, r' \in R\}$ — The server process of the server $s \in S$

$C(a) = \{\lambda \in \Lambda | \lambda = (((a, s, r), (s, v)), ((a, s', r'), (s, v'))) \vee \lambda = (((a, s, r), (s, v)), ((s, v'))), s, s' \in S, v, v' \in V, r, r' \in R\}$ — The agent process of the agent $a \in A$

Appendix C
IMDS Syntax

The IMDS syntax is presented in EBNF notation, with following rules:

Legend	
Usage	Notation
Definition	=
Non-terminal symbol	**letters and _**
Alternation	|
Optional	**[…]**
Repetition	**{ … }**
Terminal string	**" … "**
Keyword	**bold**
Range of literals	..

```
model = [title] system

title = system sys_id ";"
system = server_view | agent_view
server_view = server_types [ agent_types_list ]
         variables server_view_init
agent_view = server_headers agent_types varibles agent_view_init
server_types = { server_type }
server_type = server_header actions ";"
server_header = server_relations elements
server_relations = server ":" serv_type_id
        "(" server_formal_pars "),"
server_formal_pars = formal_pars_servers ";" formal_pars_agents |
      formal_pars_agents ";" formal_pars_servers
```

© Springer Nature Switzerland AG 2020

W. B. Daszczuk, *Integrated Model of Distributed Systems*, Studies in Computational Intelligence 817, https://doi.org/10.1007/978-3-030-12835-7

```
formal_pars_servers = servers servers_formal_pars_list
formal_pars_agents = agents agents_formal_pars_list
servers_formal_pars_list = server_formal_par { “,” server_formal_par }
agents_formal_pars_list = agent_formal_par { “,” agent_formal_par }
server_formal_par =
        server_formal_par_id [ “[” vector_size “]” ]
        [ “:” server_type_id ] |
      server_formal_par_id [ “[” vector_size “]” ] “:” “self”
agent_formal_par = agent_formal_par_id [ “[” vector_size “]” ]
        [ “:” agent_type_id ] |
        agent_formal_par_id [ “[” vector_size “]” ] “:” “self”

elements = states services | services states
states = states “{” state_list “},”
services = services “{” service_list “},”
state_list = state_def { “,” state_def }
service_list = service_def> { “,”service_def }
state_def = state_id [ “[” vector_size “]” ]
service_def = service_id> [ “[” vector_size “]” ]

actions = actions “{” action_list “},”
action_list = action { action }
action = [ condition ] [ repeaters ] input -> output “,”
condition = “?” instance_number
repeaters = repeater | repeater repeater | repeater repeater repeater
repeater = “<” repeater_id lower_limit “..” upper_limit “>”
input = “{” message “,” serv_state “}”
output = “{” message “,” serv_state “}” | “{” serv_state “}”
message = agent_term “.” server_term “.” service_term
serv_state = server_term “.” state_term
agent_term = agent_id [ “[” index “]” ]
server_term = server_id [ “[” index “]” ]

service_term = service_id [ “[” expression “]” ]
state_term = state_id [ “[” expression “]” ]

agent_types_list = agent “:” agent_types_id_list “;”
agent_types_id_list = agent_type_id { “,” agent_type_id }

variables = server_variables agent_variables |
        agent_variables server_variables
server_variables = servers server_variable_list “;”
agent_variables = agents agent_variable_list “;”
server_variable_list = server_decl { “,” server_decl }
server_decl = server_id [ “[” vector_size “]” ]
        [ “:” server_type_id ]
```

agent_variable_list = agent_decl { “,” agent_decl }
agent_decl = agent_id [“[” vector_size “]”]
 [“:” agent_type_id]

server_view_init = **init** “->” “{” <server_view_init_list> “}”
server_view_init_list = server_view_init { server_view_init }
server_view_init = server_view_server_init server_view_agent_init |
 server_view_agent_init server_view_server_init
server_view_server_init = repeater server_id “[” index “]”
 “(” server_view_parameter_list “).” server_init “,” |
 server_id “(” server_view_parameter_list “).” server_init “,”
server_view_agent_init = repeater agent_id “[” index “]” “.”
 agent_init “,” | agent_id “.” agent_init “,”
server_view_parameter_list = server_view_parameter
 { “,” server_view_parameter }
server_view_parameter = server_parameter | agent_parameter
server_parameter = server_id [“[” index_list “]”]
agent_parameter = agent_id [“[” index_list “]”]
index_list = index { “,” index } | index “..” index
server_init = state_id [“[” index “]”]
agent_init = server_id [“[” index “]”] “.”
 service_id [“[” index “]”]

server_headers = server_header “;” { server_header “;” }
agent_types = agent_type { agent_type }
agent_type = agent_header actions “;”
agent_header = **agent** “:” agent_type_id “(” formal_pars_servers “),”

agent_view_init = **init** “->” “{” agent_view_init_list “}”
agent_view_init_list = agent_view_init { agent_view_init }
agent_view_init = agent_view_server_init agent_view_agent_init |
 agent_view_agent_init agent_view_server_init
agent_view_server_init = repeater server_id “[” index “]” “.”
 server_init “,” |
 server_id “.” server_init “,”
agent_view_agent_init = repeater agent_id “[” index “]”
 “(” agent_view_parameter_list “).” agent_init “,” |
 agent_id “(” agent_view_parameter_list “).” agent_init “,”
agent_view_parameter_list = agent_view_parameter
 { “,” agent_view_parameter }
agent_view_parameter = server_parameter

system_id = identifier
server_type_id = identifier
agent_type_id = identifier
server_formal_par_id = identifier
agent_formal_par_id = identifier

```
server_id = identifier
agent_id = identifier
state_id = identifier
service_id = identifier
repeater_id = identifier
lower_limit = number
upper_limit = number
vector_size = number
instance_number = number
expression = number | repeater_id | number operator number |
        repeater_id operator number |
        number operator repeater_id
operator = "+" | "-"
number = digit { digit }
identifier = letter { character }
digit = "0" .. "9"
letter = "A" .. "Z" | "a" .. "z"
character = letter | digit | "_"
```

Changes in Timed IMDS:

```
system = server_view
server_view = server_types [ agent_types_list ]
        variables [ channels ] server_view_init
channels = channels ";" | channels "{" delays "}"
delays = delay [ "," delay ]
delay = constraint | server_ spec constraint |
        server_spec "->" server_spec constraint
server_spec = server_id [ "[" server_subrange "]" ]
server_ subrange = number | "-" number | number "%" number |
        "-" number "%" number
elements = states services [ bounds ] | services states [ bounds ]
bounds = bounds "{" bound { "," bound } "}"
bound = [ condition ] "?" clk "(" bound finishing_bracket state_id
        [ "[" index "]" ] ")"
action = [ condition ] [ restriction ] [ reset ] [ repeaters ]
        input -> [ duration ] output ","
condition = "?"
restriction = "?" clk
duration = constraint
constraint = starting_bracket number [ "," number ] finishing_bracket
starting_bracket = "(" | "<"
finishing_bracket = ")" | ">"
reset = "!"
```

Printed in the United States
By Bookmasters